T0302189

Polymer-Based Composites

Emerging Materials and Technologies

Series Editor:
Boris I. Kharissov

Polymer-Based Composites
Design, Manufacturing, and Applications

Edited by
V. Arumugaprabu, R. Deepak Joel Johnson, M. Uthayakumar, and P. Sivaranjana

CRC Press
Taylor & Francis Group
Boca Raton London New York

CRC Press is an imprint of the
Taylor & Francis Group, an **informa** business

First edition published 2022
by CRC Press
6000 Broken Sound Parkway NW, Suite 300, Boca Raton, FL 33487-2742

and by CRC Press
2 Park Square, Milton Park, Abingdon, Oxon, OX14 4RN

© 2022 Taylor & Francis Group, LLC

CRC Press is an imprint of Taylor & Francis Group, LLC

ISBN: 978-0-367-64791-9 (hbk)
ISBN: 978-0-367-64792-6 (pbk)
ISBN: 978-1-003-12630-0 (ebk)

Typeset in Times
by codeMantra

Contents

Preface

For the past few decades, composite materials have emerged as a major substitution for conventional materials in auto components, aero components, structural applications, mechanical applications, and other industries. Composites have excellent specific strength, low cost, and specific stiffness, which make them attractive alternatives for advanced technology applications. Another interesting feature of composites is its tailor-made characteristics which offer special functions for application as advanced materials. Numerous research articles, magazines, edited books, and text books have been published over the last decades narrating the fabrication and testing of different types of composites using various matrixes, fillers, and particulates. This edited book will be helpful for researchers and graduate students from various fields of engineering and science to learn specifically about polymer-based composites and their applications.

Chapter 1 discusses the history of composites and polymers, including the processing and applications of advanced composite materials. In addition, polymer-based composites and their impact on replacing other composites is also discussed. The different types of polymers used as matrixes, their properties, and their characteristics are also discussed. Along with this, an elaborate discussion on the different types of reinforcement used in the composite development is provided followed by a brief discussion on the applications of various polymer matrix-based composites.

The usage of natural fiber-based composite material is essential for developing a product for advanced applications owing to the wide availability of the materials in the nature. Chapter 2 addresses the major types of natural fiber-reinforced composites used in the development of composite materials. The different natural fibers such as sisal, banana, and pineapple, as well as their composites prepared using polymer matrixes are discussed briefly. Furthermore, the impact of synthetic fibers on the polymer matrix composites with respect to their mechanical properties is also presented. The usage of various fillers and particulates along with the polymer matrixes and their performance are also analyzed.

Chapter 3 discusses the design of various polymer-based composites, with a special emphasis on various parameters such as density, weight percentage, and voids. The chapter also presents the design of composites using various conventional manufacturing processes. The chapter further discusses the manufacturing of thermoplastic composites as well as their merits and demerits.

Another interesting concept in the development of composites is the hybrid phenomenon. The performance of hybrid composites prepared using polymers is presented in Chapter 4. Additionally, the different types of fabrication methods for hybrid composites, and their merits and demerits are discussed. Finally, the applications of hybrid composites are also discussed briefly.

Chapter 5 of this edited book discusses the modern development of composite materials, namely, biocomposites. The impact of biopolymers on the composite development is presented briefly. The various types of biocomposites involved, their processing methods, and merits/demerits are presented in the chapter.

Chapter 6 presents in detail the various nanopolymer composites developed and their special features. Chapter 7 of this edited book discusses the physical, chemical, mechanical, and environmental parameters that affect the performance of various polymer-based composites.

One of the key highlights of this edited book is the discussion on the tribo performance of the various polymer composites presented in Chapter 8. This chapter presents the wear, erosion, corrosion, and fatigue analysis of polymer-based composites with suitable case studies. Chapter 9 discusses the various failure mechanisms related to polymer-based composite materials with suitable examples. Finally, Chapter 10 elaborates the various applications of polymer-based composites in the aerospace sector, automobile sectors, and other industries with examples.

The editors and contributors of this book hope that postgraduate and undergraduate students and researchers in the field of materials science, composites, mechanical engineering, biomedical engineering, and other technologies will benefit by reading high-quality chapters related to polymer-based composite materials and their processing and applications. The research progress in this field is rapidly changing so this book is intended to be a key reference point to initiate good research in the field of composites, especially using polymer matrixes. The editors would like to acknowledge many composite researchers who have contributed to this book. The editors would also like to thank all the publishers and authors for giving permission to use their published images and original work. We hope that this book will attract more researchers to this field and form a good platform among the scientific community. Please enjoy the book and communicate to the editor/authors any comments that you might have about its content.

V. Arumugaprabu
R. Deepak Joel Johnson
M. Uthayakumar
P. Sivaranjana

Editors

Dr. V. Arumugaprabu is an Associate Professor in the Department of Mechanical Engineering, as well as Deputy Director (IQAC & CLT) in Kalasalingam Academy of Research and Education, Krishnankoil, Tamil Nadu, India. He completed his BE in mechanical engineering in 2005, ME CAD/CAM in 2007, and PhD in the field of composites in 2014. He has 13 years of teaching and 6 years of research experience. He also worked as a postdoctoral research associate in Precision Engineering Laboratory, School of Mechanical Engineering, Yeungnam University, South Korea in 2017. He has completed four PhD theses under his guidance. He has published 70 papers in reputed Scopus/SCI-indexed international journals and 40 papers in international conferences. In addition, he completed a funding project worth INR 27 lakhs from MOEF, India. He serves as an editorial board member for three journals. He has also received five awards for teaching and research competence from IQAC, KARE. He is a lifetime member of ISTE, InSc, and ISRD. His areas of interests include composite materials, industrial safety engineering, and machining studies.

Dr. R. Deepak Joel Johnson is an Assistant Professor in the Department of Mechanical Engineering and has a strong knowledge of polymer composites and biocomposites. He completed his full-time PhD in mechanical engineering from Kalasalingam Academy of Research and Education (Deemed to be University), Virudhunagar, Tamil Nadu, India in 2019. He completed his BTech in mechanical engineering with first class in May 2012 and MTech in mechanical engineering specialized in manufacturing with first class in May 2014 from Karunya University, Coimbatore, Tamil Nadu, India. He has more than 6 years of teaching and research experience. He has published more than 25 papers in international journals and two book chapters with reputed international publishers.

Dr. M. Uthayakumar was born in Cumbum, India and obtained his Master of Engineering in Production Engineering from Thiagarajar College of Engineering (Autonomous), Madurai, India. He conducted research on bimetallic piston machining studies in the Department of Production Engineering, National Institute of Technology, Tiruchirappalli. He was awarded as a young scientist by the Tamil Nadu state council for Science and Technology and completed a postdoctoral fellowship on Tribology in the Department of Mechanical Engineering, Indian Institute of Technology, New Delhi. He has published 140 papers in international journals. He has visited Cracow University of Technology, Poland and Yeungnam University, Korea as a visiting professor. He organized the second international conference on Advanced Manufacturing and Automation (INCAMA 2013) successfully and produced special issues in reputed journals. He successfully completed a research project for the Department of Atomic Energy-Board of Studies in Nuclear Science (DAE-BRNS). Currently, he is involved in research projects funded by DST, DRDO, MoEF, and CC, as well as an international research project by Poland National Science Academy. He has visited Sri Lanka, Malaysia, Poland, Dubai, Abu Dhabi, Sarjah, Germany, South Korea, Hungary, and Slovakia. Ten of his PhD scholars have completed their studies and eight more are pursuing. He serves as the National Executive Committee member of ISTE, New Delhi and a Management Committee member of SAEINDIA Southern section. He received outstanding scientist award by VDGOOD Technology Factory in 2020. Currently, he is working as a Professor in the Faculty of Mechanical Engineering, Kalasalingam Academy of Research and Education, Krishnankoil, Virudhunagar, Tamil Nadu, India.

P. Sivaranjana has been working as an Assistant Professor in chemistry for about 17 years and has a research experience of about 5 years. She completed her undergraduation and postgraduation in chemistry at Standard Fireworks Rajaratnam College for women, Sivakasi, under Madurai Kamarajar University. She completed her PhD at Kalasalingam Academy of Research and Education in 2019. Her area of research is biocomposites. She has developed three different kinds of biodegradable packaging materials from cellulose using agro waste and two different kinds of fabric composites for biomedical applications. She has more than 10 international publications.

Contributors

P. Amuthakkannan
Department of Mechanical Engineering
PSR Engineering College
Sivakasi, Tamil Nadu, India

K. ArunPrasath
Department of Mechanical Engineering
Kalasalingam Academy of Research
and Education
Krishnankoil-626 126, Tamil Nadu,
India

N.C. Brintha
Department of Computer Science &
Engineering
Kalasalingam Academy of Research
and Education
Krishnankoil-626 126, Tamil Nadu,
India

Oisik Das
Department of Engineering Sciences
and Mathematics
Luleå University of Technology
Luleå 97187, Sweden

Onkar A. Deorukhkar
Dr. Vishwanath Karad MIT World
Peace University
Paud Road, Kothrud, Pune-411 038,
Maharashtra, India

Prachi Desai
Dr. Vishwanath Karad MIT World
Peace University
Paud Road, Kothrud, Pune-411 038,
Maharashtra, India

Aravind Dhandapani
Department of Mechanical Engineering
Kalasalingam Academy of Research
and Education
Krishnankoil-626 126, Tamil Nadu,
India
and
University Science Instrumentation
Centre
Madurai Kamaraj University
Palkalai Nagar, Madurai-625 021, Tamil
Nadu, India

C. Pradeepkumar
Department of Automobile Engineering
Kalasalingam Academy of Research
and Education
Krishnankoil-626 126, Tamil Nadu,
India

Senthilkumar Krishnasamy
Center of Innovation in Design and
Engineering for Manufacturing
(CoI-DEM)
King Mongkut's University of
Technology North Bangkok
1518 Wongsawang Road, Bangsue,
Bangkok 10800, Thailand

N.B. Karthikbabu
Centurion University of Technology and
Management
R. Sitapur, Odisha, India

Malhari Kulkarni
Dr. Vishwanath Karad MIT World
Peace University
Paud Road, Kothrud, Pune-411 038,
Maharashtra, India

M. B. Kulkarni
Dr. Vishwanath Karad MIT World
  Peace University
Paud Road, Kothrud, Pune-411 038,
  Maharashtra, India

K. Mayandi
Department of Mechanical Engineering
Kalasalingam Academy of Research
  and Education
Krishnankoil-626 126, Tamil Nadu,
  India

Yashwant S. Munde
MKSSS's Cummins College of
  Engineering for Women
Karvenagar, Pune-411 052,
  Maharashtra, India
and
Savitribai Phule Pune University
Maharashtra, India

Chandrasekar Muthukumar
School of Aeronautical Sciences
Hindustan Institute of Technology &
  Science
Padur, Kelambakkam, Chennai-603
  103, Tamil Nadu, India.

S. Radhakrishnan
Dr. Vishwanath Karad MIT World
  Peace University
Paud Road, Kothrud, Pune-411 038,
  Maharashtra, India

S. Rajesh
Associate Professor, Department of
  Mechanical Engineering
Kalasalingam Academy of Research
  and Education
Krishnankoil-626 126, Tamil Nadu,
  India

Sundarakannan Rajendran
Faculty of Mechanical Engineering
Kalasalingam Academy of Research
  and Education
Krishnankoil-626 126, Tamil Nadu,
  India

N. Rajini
Department of Mechanical Engineering
Kalasalingam Academy of Research
  and Education
Krishnankoil-626 126, Tamil Nadu,
  India

N. Sabarirajan
Chendhuran College of Engineering and
  Technology
Pudukkottai, Tamil Nadu, India

K. Sankaranarayanan
Department of Mechanical Engineering
Kalasalingam Academy of Research
  and Education
Krishnankoil-626 126, Tamil Nadu,
  India

Dr. T. Sathish
Saveetha School of Engineering,
  SIMATS
Chennai 602 105, Tamil Nadu, India

Vigneshwaran Shanmugam
Faculty of Mechanical Engineering
Saveetha Institute of Medical and
  Technical Sciences
Kuthambakkam, Tamil Nadu, India

Sanam Shikalgar
Dr. Vishwanath Karad MIT World
  Peace University
Paud Road, Kothrud, Pune-411 038,
  Maharashtra, India

Suchart Siengchin
Department of Materials and Production
  Engineering, The Sirindhorn
  International Thai-German Graduate
  School of Engineering (TGGS)
King Mongkut's University of
  Technology North Bangkok
1518 Wongsawang Road, Bangsue,
  Bangkok 10800, Thailand

P. Sivaranjana
Department of Chemistry
Kalasalingam Academy of Research
  and Education
Krishnankoil-626 126, Tamil Nadu,
  India

P. Sivasubramanian
Department of Mechanical Engineering
Kalasalingam Academy of Research
  and Education
Krishnankoil-626 126, Tamil Nadu,
  India

Karthikeyan Subramanian
Department of Automobile Engineering
Kalasalingam Academy of Research
  and Education
Krishnankoil-626 126, Tamil Nadu,
  India

Senthil Muthu Kumar Thiagamani
Department of Automobile Engineering
Kalasalingam Academy of Research
  and Education
Krishnankoil-626 126, Tamil Nadu,
  India

M. Thirukumaran
Department of Mechanical and
  Automation Engineering
PSN College of Engineering and
  Technology
Melathediyoor, Tirunelveli, Tamil
  Nadu, India

S. Vignesh
Department of Mechanical Engineering
Kalasalingam Academy of Research
  and Education
Krishnankoil-626 126, Tamil Nadu,
  India

J.T. Winowin Jappes
Department of Mechanical Engineering
Kalasalingam Academy of Research
  and Education
Krishnankoil-626 126, Tamil Nadu,
  India

1 History of Composites and Polymers

P. Sivasubramanian, K. Mayandi,
V. Arumugaprabu, N. Rajini, and S. Rajesh
Kalasalingam Academy of Research and Education

CONTENTS

1.1 DEFINITION OF COMPOSITES

A new type of engineering material emerged in the mid-20th century viz., composite materials. Two or more things with different behaviors combining to form a new thing are termed composites. Similarly, "composite materials" are formed by combining two materials with different properties. Figure 1.1 shows some examples of the combination of two materials.

Historically, the term compositeis not new as its usage existed in the ancient ages. A few good examples are: bricks made by ancient civilizations were composed of mud and straw,the ancient Japanese-made Samurai swords, Damascus gun barrels made using layers of iron and steel, and mud walls built using bamboo shoots. Progress results in technological developments, which mainly depends on innovations in materials. In this sense, applications of composite materials are gradually increasing, replacing conventional materials. The excellent merits offered by composite materials, such as low weight, high strength, and availability, make it as leader among the various types of materials used. Furthermore, composites may be of natural origin or industrially produced.

Nature itself has some good examples, mainly from animal and plant sources, where composite materials are used. Composites consist of two phases, namely, the matrix phase and the reinforcement phase. Matrix is the primary phase having a continuous character, and reinforcement is the secondary phase having a discontinuous form. Some matrices are usually more ductile and less rigid, consisting of polymers, ceramics, or metals as one of the three major material types. In composite materials, the matrix is the main component, and the reinforcement phase is discontinuously

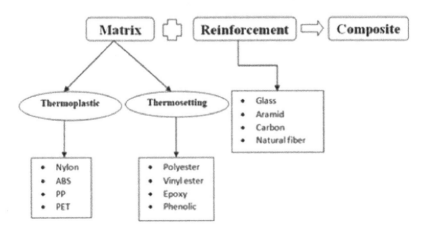

FIGURE 1.1   Combinations of two materials.

integrated into the matrix. The dispersed phase is usually more complicated than the continuous phase and is referred to as reinforcement. It enhances composites and improves the mechanical properties of the matrix. The three familiar types of matrices involved in the production of composites for various applications include polymer matrix, metal matrix, and ceramic matrix-based composites.

The following three types of reinforcements are commonly used: fibers, particulates, and laminates. Figure 1.2 shows the family tree of composites. The term composites as it refers to composite materials is defined as *"Two or more materials combined on a macroscopic scale to form a third new material which possesses certain special property characteristics."*

Forming composite materials, leads to improvement in characteristics such as weight, strength, stiffness, wear resistance, corrosion resistance, electrical resistance, acoustic enhancement, thermal property, temperature resistance, moisture resistance, chemical resistance, and fatigue. Currently, many composites are developed based on the environmental aspect so that usage of wastes as filler or reinforcement and matrices leads to tremendous improvement in the properties. Moreover, the concept of biocomposites is attracting the world in such a way that complete degradation is possible. The more important thing is that we should not consider all two or more materials as composites, as a composite material must satisfy three essential criteria as given below:

1. The two materials combining should be able to fabricate.
2. It needs to have two or more physically or chemically distinct phases with an interface separating the two materials.
3. The final material property obtained must be a new property that does not reflect any one of the specific material properties among the two.

Our focus in this book is on the fabrication of polymer matrix composites (PMCs) using natural fiber reinforcements, as well as its mechanical properties, wear properties, erosion properties, and fatigue properties. Furthermore, the development of

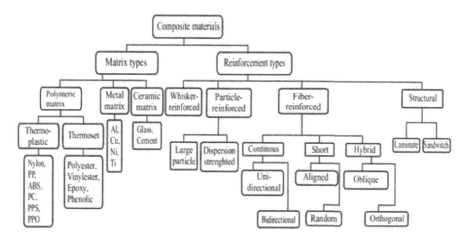

FIGURE 1.2   Family tree of composites.

nanocomposites, biocomposites using the different types of polymer matrices, hybrid composites, and the failure analysis of composites are discussed in detail. Finally, a discussion on the various applications of PMCs is also presented.

1.2 DEFINITION OF POLYMERS

The term polymer is not new as it has been used since the ancient period. Polymers are made from pieces (known as monomers) that can be easily connected into a long piece similar to a chain (known as polymers). Let us begin with monomers. They are low-molecular-weight compounds that combine to form a polymer, which a high-molecular-weight molecule composed of small repeated units. According to the International Union of Pure and Applied Chemistry, *"A polymer is a substance composed of molecules characterized by the multiple repetitions of one or more species of atoms or groups of atoms (constitutional repeating units) linked to each other in amounts sufficient to provide a set of properties that do not vary markedly with the addition of one or a few of the constitutional repeating units."*

Naturally occurring polymers used for centuries include wood, cotton, leather, rubber, wool, and silk. An excellent example of a human made-made polymer is nylon. Polymers are broadly classified into the following four types:

(a) Thermoset – A cross-linked polymer not able to melt, for example, rubbers and tires.
(b) Thermoplastic – A cross-linked polymer that can melt, for example, plastic.
(c) Elastomer – A polymer that undergoes deformation, that is, stretch and returns to its original form. It is also known as thermoset polymers.
(d) Thermoplastic elastomer – An elastic polymer that can melt, for example, footwear and sole of shoes.

In addition to the above, other polymer families include (i) polyolefin, made from monomers linked with olefin; (ii) polyesters, amide, urethanes, made from monomers linked with ester, amide, urethanes, and other functional groups; and (iii) natural polymers such as DNA, proteins, and polysaccharides.

Figure 1.3 shows the broad types of natural and synthetic polymers used widely in various applications. The key point that makes polymers unique is that they comprise larger molecules or macromolecules that give new properties compared to the smaller monomers. Another excellent feature that makes polymers unique is that the chain mesh, longer polymer chains meshed, provides more flexibility. Polymers are produced using two common mechanisms, polymerization and step-growth polymerization. In addition to polymerization, polymers are manufactured by sequentially adding monomers using reaction; in which the growth of the polymer chain is linear. On the other hand, in step-growth polymerization or condensation polymerization, monomers initially react together and form small oligomers, which then lead to the formation of polymers. Another process of manufacturing polymers is by condensation reaction including monomers containing two functional groups, in which even small molecules like water are eliminated. An excellent example of a polymer manufactured using a condensation reaction is nylon. In addition, the copolymers concept plays a significant

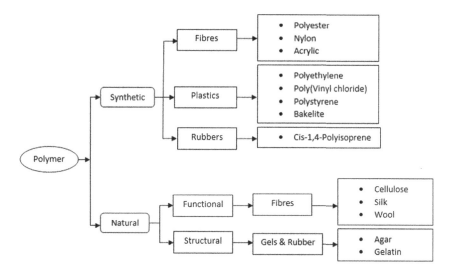

FIGURE 1.3 Types of natural and synthetic polymers.

role in producing polymers. Copolymers are composed of more than one monomer; an excellent example of this is proteins formed by combining many amino acids. The condensation polymer and copolymer have a mutual relationship with each other with respect to their function. All natural polymers are copolymers. The polymer structure depends on its molecular shape and mass. As the polymers have long chains, the molecular weight plays a vital role in the structure of the polymer. Polymers have a high molecular mass. Moreover, the polymer chain can bend more flexibly, making it capable of producing different shapes with unique characteristics. The two essential physical properties of polymers are strength and flexibility, which depend on the various parameters such as chain length, side groups, branching, and cross-linking. The polymer structure is divided into the following categories:

1. Linear polymers
2. Branched polymers
3. Cross-linked polymers
4. Network polymers

Linear polymers are single flexible chains in which monomers are joined end to end. Examples of linear polymers include polyethylene, polyvinyl chloride, polystyrene, and nylon. Branched polymers have branched chains connected to the main chain. The branches, considered part of the main chain, are formed during side reactions during the synthesis of polymers. Cross-linking is achieved either during synthesis or by a nonreversible chemical response usually carried out at an elevated temperature. Rubber is an example of a cross-linked polymer. Network polymers have trifunctional monomer units with distinct mechanical and thermal properties. Examples of network polymers include epoxies and other Adhesives.

Figure 1.4 shows the familiar sources where polymers are effectively used.

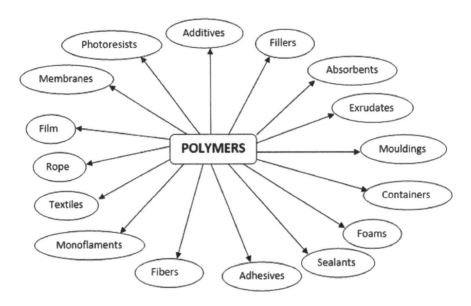

FIGURE 1.4 Common application and related geometries of polymers.

Merits:

1. They are nontoxic.
2. They possess less weight.
3. They have good mechanical strength.
4. They have good moisture and corrosion resistance.
5. They are easy to fabricate and safe to use.

Demerits:

1. Some polymers easily burn.
2. Polymer functions and properties vary, with some of them not resistant to high temperatures.

1.3 NEED FOR POLYMER-BASED COMPOSITES

A polymer composite comprises a variety of short or continuous fibers or fillers bound together by an organic polymer matrix. Polymers make ideal matrix materials as they can be processed quickly, are lightweight, and have desirable mechanical properties. Based on their structure, polymers are more complex than metals and ceramics. Excellent resistance to chemicals offered by polymers is comparable to that of metals. However, compared to metals, polymers cannot withstand high temperatures. Additionally, polymers are not a good conductor of electricity. Despite these demerits, use of polymer-based composites in various industrial applications has increased because of their performance.

One of the critical aspects is the development of various high-performance natural/synthetic fibers as reinforcements along with polymers. Some examples of natural fibers include banana, sisal, pineapple, bamboo, carbon, aramid, glass, basalt fiber, etc. The polymer matrices show some excellent performance enhancement in mechanical properties, thermal properties, and electrical properties along with these reinforcements. The fibers and polymers mixed in varying proportions create a new material with excellent performances, satisfying different requirements. In addition, along with fibers, multiple types of fillers such as $CaCO_3$, clay, mica, and glass microspheres reused as reinforcements, with polymers displaying an improvement in property.

The addition of fillers in the PMCs reduces cost, reduces shrinkage of mold, increases modulus, and enables a smooth surface. The impact strength of composites significantly improves by adding fillers along with the polymers, as well as the crack resistance. The need for polymer-based composites is crucial as it is as an effective engineering material used for high-end laboratory purposes and low-cost household items such as handbags, chairs, tables, and sports goods.

PMCs are utilized to transfer loads between the fibers or fillers of a matrix. A few of the major benefits with PMCs include lightweight, high stiffness, and high strength of the reinforcement. These excellent advantages lead to PMCs application in aircraft, automobiles, and other structural applications. Other essential properties such as corrosion and fatigue resistance are also high for PMCs compared to metals. Based on research conducted over the last 15 years has shown that advanced composite structures using PMC can be used to fabricate military aircrafts.

1.4 POLYMER MATRIX TYPES

The two main types of polymers used as matrix materials are thermosets and thermoplastics. The more commonly used thermoset polymers and their applications include the following:

(i) Epoxies, polyesters, and vinyl esters used in aerospace, automotive, marine, and chemical applications.
(ii) Phenolics: used in bulk molding compounds.

The more commonly used thermoplastic polymers and their applications include the following:

(i) Nylons (such as nylon 6, nylon 6.6), thermoplastic polyesters (such as polyethylene terephthalate [PET], PBT) used with discontinuous fibers in injection-molded articles.
(ii) Polyamide-imide (PAI), polyether ether ketone (PEEK), polysulfone (PSUL), polyphenylene sulfide (PPS), and polyetherimide (PEI) suitable for moderately high-temperature applications with continuous fibers (Mallick, 2008).

1.4.1 Thermoset Polymers

Thermoset polymers are also known as thermosetting plastic or thermosetting polymers. Thermosetting polymers are permanently hard that do not soften when heated. They are composed of a network of polymers in which the covalent bond resists motion at high temperatures. The commonly used thermoset polymers for production of composites include polyester, epoxy, vinyl ester, phenolic resins, polyamides, and bismaleimides. Among the available matrices, the thermoset polymers play a crucial role in the development of advanced materials. For the past 20 years in composites development, polyester, epoxy, and vinyl ester resins are utilized in synthesizing composites for various engineering applications. An advanced composite for the aerospace and automobile sectors has been developed using the epoxy resin. These thermosets are used rapidly because of their three-dimensional cross-linked structure, which possesses unique features such as resistance to temperature and solvents.

Merits:

1. They are more cost-effective.
2. They allow high design flexibility.
3. They are more resistant to high temperatures.
4. They have high dimensional stability.

Demerits:

1. Recycling and reuse of the products.
2. Products cannot be reshaped easily because of the mold.
3. Surface finish is not sufficient.

1.4.1.1 Polyester

Polyesters area type of thermoset polymer in which the main chain consists of an ester functional group. Polyesters are commonly used polymers for composite production, and are generally made up of a chemical called ester, which must be repeated. Polyester, as a specific material, belongs to the type of PET. Another exciting thing about polyester is that depending on the chemical structure, it can also be a thermoplastic. Polyesters can be manufactured by natural and synthetic techniques. Synthetic polyester is manufactures through the step-growth polymerization process, whereas natural polyesters are manufactures, using plants in such a way that composition such as lipid and hydrocarbon polymers are impregnated with wax known as plant cuticles. Natural polyesters are more efficient which makes them biodegradable, while few synthetic fibers are biodegradable. However, most polyester are nonbiodegradable, the popularity of synthetic polyesters is higher. Polyesters have low moisture absorption, low weight, flexibility, long life, strength, etc. Furthermore, polyester resins are hydrophobic and resistant to minerals; organic acids also have a suitable melting point.

Among the market share in the use of polymers, polyesters occupy an 18% share of use in various products. Bottles, film, canoes, liquid crystal, holograms, movie films, Wire film insulation, and insulating tapes can be manufactured using polyesters. Additionally, they are widely used in guitars, pianos, and vehicle/yacht interiors.

1.4.1.2 Epoxy

Epoxies are thermoset resins used for making advanced composites, and carbon fiber and glass fiber attract more applications in the field of aircraft and automobile sectors. Epoxy resin is a very famous type of resin used for various industrial applications since 1946. Similar to polyester, epoxy has an aromatic group that makes it more robust, resulting in superior performance. This also makes epoxies more thermostable. Commonly used epoxies have a low molecular weight, that is, oligomers producing using step-growth polymerization reaction or condensation polymerization that combine bisphenol-A epichlorohydrin. Along with this,for the curing of epoxy resin, amines and anhydrides are used. The following advantages make epoxy excellent compared to polyester and vinyl ester:

1. They have a better bonding property, which offers them high specific strength.
2. They possess very high moisture resistance.
3. They have high durability in such a manner that their weight does not vary throughout their life period.
4. They possess good electrical resistance, thermal resistance, and corrosion resistance.
5. Their shrinkage rate is low during the curing process.

Epoxy resins are widely used in various engineering applications such as coatings, laminates, panels, tooling, molding, construction, aerospace, and automobile applications. The drawbacks of epoxy include high cost, high viscosity, and more brittle.

1.4.1.3 Vinyl Ester

The third category of thermosets is vinyl ester resin in the intermediate range compared to polyester and epoxy in terms of cost and mechanical properties. Vinyl ester resin is usually classified as unsaturated polyester resins. It follows the same processing method and curing as polyester, but its mechanical properties are similar to that of epoxy. They are high-performance resins formed by mixing epoxies with unsaturated carboxylic acids. They offer good corrosion resistance and good physical properties compared to polyester resins. They are used as excellent coating and adhesive materials. Some of the applications of vinyl ester along with fiber-reinforced composites are in the paper and pulp industries, chemical storage tank, pipes, automobile parts, optical fiber coatings, and bridges.

1.4.1.4 Phenolic Resins

Phenolic resins are the first synthetic polymer resin formed using a reaction involving phenol with formaldehyde. This type of resins is used especially when there is a requirement for fire and temperature resistance. They are also used as a binder in manufacturing various household items and natural fiber reinforcements used for various structural applications.

1.4.1.5 Polyamides and Bismaleimides

The next type of polymer is polyamides in which the repeating units of polymer are joined by amide bonds. Two types of polyamides are used, namely, natural and synthetic ones. Examples of naturally occurring polyamides are proteins and synthetic polymers used in automobile sectors, sportswear, and carpets. Another essential type of polymer used in advanced composites is bismaleimides, which is manufactured using a diamine reaction with maleic anhydride. Their manufacturing process is the same as that of epoxy. These five types of thermoset polymers are used in various applications.

1.4.2 THERMOPLASTIC POLYMERS

Thermoplastic polymers are made up of a lengthy cross-linked chain of molecules. They are also called thermosoftening plastic. The thermoplastic structure is rigid and utilizes heating the thermoplastics to provide different shapes, which are molded and pressed. After heating and pressing, they are permanently set. The excellent feature of thermoplastic polymer is that it can change its shape and can be remolded. A typical thermoplastic process cycle is shown in Figure 1.5.

Similar to thermosets, thermoplastics also have many types of resins in use for various applications. Each type of resin involved for typical low-to-high stress applications depends on the requirement.

1.4.2.1 Thermoplastic Properties

Thermoplastics are generally high strength, flexible, and shrink resistant according to the type of the resin. They are versatile materials, ranging from plastic carrier bags to increased stress and mechanical precision parts. These materials are suitable for all applications.

1.4.2.2 Thermoplastic Processing

Thermoplastics can be manufactured using injection molding, thermoforming, extrusion molding, and vacuum forming. Granular material, usually spherical granules around 3 mm in diameter, is fed into the mold. Then these granules are heated until the melting point is very high.

Because thermoplastics are highly efficient thermal insulators, they take longer to cool than other plastics during the curing process. Quick refreshment usually occurs by spraying with cold water or plunging into water baths to achieve high output rates. Cold air is blown on the surface to cool thermoplastic plastic films. The plastic

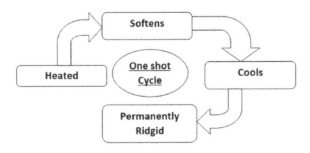

FIGURE 1.5   Thermoplastic process cycle.

shrinks after refrigeration, depending on the material and varies between 0.6% and 4%. The cooling and retraining rate has a distinctive effect on the material and the internal structure crystallization, and therefore, the retraction rate for thermoplastics is always determined.

1.5 TYPES OF POLYMER-BASED COMPOSITES

Composite matrices are usually either thermosets or thermoplastics. Nowadays, thermosets are utilized as the predominant types. Thermosets are sub-divided into multiple resin systems as polyester, epoxy, phenolic, polyurethane, and polyimide. Epoxy systems currently dominate the advanced composite industry. Polymer composite materials depend on the type of components shown in Figure 1.6.

1.5.1 MATRIX PHASE

1.5.1.1 Polymer Matrix Composites (PMC)

PMCs consist of thermosetting plastics or thermoplastics matrix with scattered carbon, glass, Kevlar, and metal fibers. Due to their higher strength and resistance to high temperatures, thermosets are more common than thermoplastics. Mixing resin and hardener prepares thermosets. The laminar structure is most commonly achieved by stacking and combining thin fiber and polymer layers until the desired thickness is achieved. PMCs are low-cost composites due to fast handling and quick manufacturing methods.

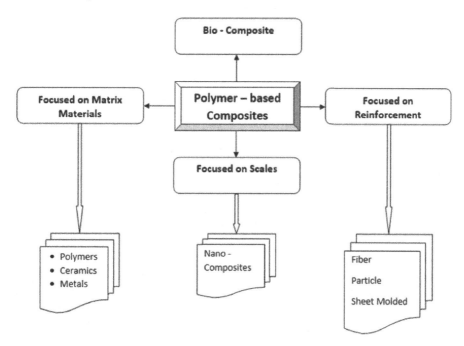

FIGURE 1.6   Types of polymer-based composite materials.

1.5.1.2   Ceramic Matrix Composites (CMC)

CMCs are generally composed of ceramic carbon, carbide silicone (SiC), aluminum oxide (Al_2O_3), and silicon nitride fibers embedded into the ceramic matrix structure. They are designed to resolve monolithic ceramics and brittleness. Due to failure, matrix strain is lower than fiber failure strain; CMCs are referred to as inverse composites. However, this phenomenon is reverse in most polymer or metal composites. Therefore, under loading conditions, the matrix fails first to avoid early failure of brittle fibers. CMCs are manufactured using specific processing techniques called gas or liquid-phase routes. Through this cycle, gaseous or liquid precursors form around the fibers with the inner phase and matrix.

1.5.1.3   Metal Matrix Composites (MMCs)

MMCs are generally composed of ceramic carbon, SiC, Al_2O_3, and silicon nitride fibers embedded into the ceramic matrix structure. They were designed to overcome the inconvenience and fragility of monolithic ceramics. The MMC matrix material is metallic (mostly aluminum, magnesium, copper, and titanium), and reinforcements can either contain scattered or metallic ceramics (e.g., tungsten, molybdenum, lead). Strengthening contributes a few percentage points to around 50% of the composite material's total volume. Al-based MMCs were widely used by reinforcement compounds such as SiC and Al_2O_3in the automotive and aerospace industries. They can be mixed quickly and effectively in molten Al to achieve desired characteristics such as superior strength, improved stiffness, reduced density, controlled thermal expansion, and better wear resistance. Due to the high rigidity and abrasive structure, the tool rate during MMC machining is highly unconventional, where in there is no contact between the tool and the material. MMCs usually utilize unique machinery techniques.

1.5.2   REINFORCEMENTS

1.5.2.1   Fiber

The spreading phases of synthetic fibers such as glass, carbon, basalt, and Kevlar in a composite structure provide enhanced material properties such as high strength, rigidity and chemical resistance, temperature resistance, and wear resistance in reinforced composite materials. The use of natural fiber enhancement is currently becoming enormously popular among scientists. Chemically treated natural fibers that show improved impact and fatigue strength are available at low cost, are biodegradable and therefore environmentally friendly, and have low density compared with synthetic fibers.

1.5.2.2   Particle

Particle-based composites are less effective with respect to strength compared to fiber-based composites. Enhanced particles find applications where high wear resistance, such as road surfaces, is required. By adding gravel as reinforcement filler material, the hardness of cement significantly increases. Low-cost and ease of production and formation are advantageous to particulate-reinforced composites. Concrete is an excellent example; the coarse rock or gravel aggregation is embedded in a cement matrix. Here, aggregates provide rigidity and strength as cement binding serves to hold the structures together. When high-volume fractions of glass-based iron particles strengthen MMCs, the Al matrix is substantially hardened, resulting in an effective combination of high strength and plasticity results. The aggregate also has high strength.

1.5.2.3 Sheet

The sheet-based composite, commonly referred to as the sheet molding composite (SMC), is a glass-reinforced, thermoset molding material, usually molded using the compression method. It incorporates long fiberglass and unsaturated resins to make a composite for high strength molding. For large structural parts, SMCs are applicable as they show a high strength-to-weight ratio. They offer additional advantages in the part design such as fasteners and subsets. The buckling temperature is higher for the spherical panel compared to the cylindrical coating panel.

1.5.3 SCALE

1.5.3.1 Nanocomposites

The desired material properties can be achieved using two or more distinct nanoscale materials, which lead to the forging of new nanocomposite material. Nanocomposites are generally classified as unintercalated, intercalated, and exfoliated composites and are produced using template synthesis, polymer intercalations, polymerization in situ, and melting. Nanocompounds for biomedical applications such as dental therapy, bone tissue technology, drugs in cancer therapy, and lesions are biomedical nanocomposites.

The integration of transparent matrix material efficiently uses exceptional optical properties of composite materials. Graphene carbon nanotubes and nanocomposites MoS_2/graphene show advantageous optoelectronic features for photonic use, while nanostructure for black phosphorus is used in biomedical applications such as cancer.

1.5.4 BIOCOMPOSITES

Researchers are looking for new opportunities to develop biocomposites as demands for biodegradable, environment-friendly materials have increased. Biocomposites made of palm sugar, enhanced in the sago starch matrix, show improved thermal stability, decreased water absorption, and increased tensile strength, tearing strength, and material durability. In various applications, such as tissue engineering scaffolds, and biopackaging, bio-nanocomposites produced at nanoscales can demonstrate potential. To sustain the quality of food packaging, the antibacterial property of ginger fibers has been effectively used.

1.6 TYPES OF REINFORCEMENTS

Reinforcements add rigidity as well as significantly prevent crack spread. The matrix's mechanical characteristics are remarkably stronger, harder, and more rigid than that of the matrix in most cases. Reinforcements are divided into four fundamental categories: flakes, fillers, particulates, and fibers.

Flakes are flat and mainly have two-dimensional geometry with two directions of strength and steepness. When suspended in a glass or plastic, they can form a useful composite material. Usually, flakes are packaged with a higher density than fiber packing concepts in parallel with each other. Mica, aluminum, and silver are typical flake materials. Mica flakes integrated into a glassy matrix provide easy-to-work composites for electrical use. In paints and other layers, aluminum flowers are commonly used for orientation to the coating surface. When good conductivity is required, silver flakes are used.

Fillers are particles or powders added to the material to change and improve the composites' mechanical and physical properties. They also serve to reduce the consumption of a more expensive binder. Fillers are used to modify or improve thermal conductivity, electric resistivity, friction, wear resistance, and flame resistance. Calcium carbonate, aluminum oxide, chalk, fumed silica, treated earth, and hollow glass beads are commonly referred to as calcium oxide.

Small particles ($<0.25\,\mu m$), hollow spheres, cubes, platforms, or carbon nanotubes are used in composites. The particles have desirable material properties in each case, and the matrix is the binding medium required for structural applications. The layout can be random or with a preferred orientation of the particulate material. In general, particles do not improve strength and resistance to fractures very efficiently. Typical materials include lead, tungsten, copper, chromium, and molybdenum.

Finally, fiber is a rope or string used as an element of composite materials, which are usually very large (>100) in terms of aspect ratio (length/diameter). The transversal section can be circular, hexagonal, or square. The following are commonly used fibers composites: glass fiber consisting mainly of silicon and metal oxide-modifying elements and are generally produced through a small orifice through the mechanical drawing of molten glass. They are used extensively because of low cost and high corrosion resistance. Fiberglass is stronger and lighter than glass sheets and more ductile than carbon. The aramid fiber can be used as a fiber in fishing rods, storage pans, and aircraft parts. Examples of industrial application are armor, clothes for protection, and sports products. Carbon fibers are frequently produced using a polyacrylonitrile oxidized or carbonized polymer pyrolysis. Carbon fiber may have a low-density modulus of up to 950 GPa. It usually has a diameter of between 5 and 8 mm, less than a human hair (50 mm) and boron fiber, which generally have high rigidity, good input strength, and large (0.05–0.2 mm) diameter compared with other types of fiber. Composites with boron fibers are commonly applied in aerospace structures where high rigidity is required, and silicon carbide fibers are used because of their excellent oxidation resistance. Moreover, it has superior modulus and resistance in high-temperature atmospheres, which are generally used in high-temperature metallic and CMCs.

1.7 APPLICATIONS

1.7.1 POLYMERS COMPOSITES IN THE TRANSPORTATION INDUSTRY

The transportation industry, broadly comprising automotive, aerospace, and marine sectors, utilizes polymer composites in various components. Some of the more common uses are summarized below.

Automotive vehicles: The timing belt used in Toyota Camry (1993) was the first component polymer matrix nanocomposite used in a commercial automobile product. Other applications have followed this breakthrough over the decades, such as bumpers, board panels, fuel tanks, mirror boxes, and engine parts. The technology is currently expanding with composites finding more and more applications in automobiles.

- Polymer composite matrix is used in tires along with a few varieties of belts. The tires are made up of rubbers with different additives. Carbon black, which improves strength and durability, is the most important of these additives.
- Fiber-reinforced polymer (FRP) composite components are used in automotive bodies (external body panels).
- A few expensive sports car makers such as Bugatti use carbon fiber-enhanced PMC to construct the vehicle's entire external body.
- Other applications such as fins, bumpers, slats, body panels, and engine components followed this breakthrough over the decades. The technologies have stretched to decrease tire-rolling resistance and provide paint using ultra-hard coatings, glass for the windscreen, and light bulbs.
- Indoor panels, instrument panels, headlight assembly, taillight assembly, and trims can also be manufactured using polymers and polymer composites.
- Textiles are made of polymeric fibers, with applications asseat covers, seat belts, airbags, and carpets. Fabrics used for car interiors have historically been derived from dominant fossil fuel feedstocks. However, with sustainability becoming important, textiles based on biopolymers are frequently used as options for petroleum-based plastics in automotive interiors.
- The favored material for filling the internal part of automotive seats is flexible polyurethane foam.

Aerospace vehicles: PMCs have been commonly used in tires and interiors. The aviation industry's relentless drive to increase performance by reducing weight and advancements in the material properties of composites has increased their popularity. Fiber-infused PMCs can, most importantly, be optimized to combine high resistance, rigidity, toughness, and low density. This combination of properties helps in achieving exceptional strength-to-density and rigidity-to-density ratios, along with superior physical properties. This often makes them the preferred structural materials for use in aircraft components.

- Composites of the polymer matrix are also used in aircraft tires and indoor applications.
- Aircraft tires need to meet even higher performance and reliability than car tires, as they must withstand higher loads for short durations at landing.
- Aircraft windows are made of lightweight polymers with relatively healthy optical features, such as polycarbonate or acrylic plastic.
- Polymers and PMCs are used as construction materials in interior components of aircraft, such as interior panels, instrument panels, tabletops, bar tops, countertops, doors, cabinets, trim, casings, and overhead storage bins.

- Many interior components such as seat coverage, seat belts, and taping are used for clothing textiles made from polymeric fibers.
- A varied type of polymeric foam is used as seat covers for applications as flame-retardant memory foam, closed-cell polyethylene foam, flame-retardant polyurethane, high-resilience flame-retardant polyurethane foam, and open-cell silicone foam.

Marine vehicles: Composites of polymer matrix find several applications in marine vehicles. Fiberglass boats are some of the best-known examples as fiberglass is a composite, in which glass fibers can be randomly arranged or chopped as a strand mat or a fabric-reinforced matrix polymer. Instead of glass fibers, the increasing use of light, stiffer, and stronger carbon fibers is an emerging trend in the construction of boats.

- The term "boat" refers commonly to any size and type of marine vehicle. This range of vehicles include canoes, fishers, yachts, passenger boats, cruise liners, warships, and aircraft carriers.
- Polymers and PMCs are used as the primary construction material in the instrument panel, box, tabletop, bar top, countertops, doors, cabinets, and trim.
- The primary building materials for some boat shells are PMCs.
- Many interior components, including seat covering and carpeting, are used in textiles built of polymeric fiber.
- Many kinds of polymer materials are used in boat seat coils: compressed polyester, polyester fiberfill, medium-density antimicrobial polyurethane foam, high-density polyurethane foambulant.
- The inherent, intrinsic life jackets were made using natural materials such as cork, balsa wood, and kapok; however, today, most frequently used materials are synthetic polymer foams such as PVC or polyethylene.
- A long-lasting fabric made of nylon or polyester to withstand extended contact with water is an outside shell of a life jacket.
- Sailcloth is produced using various suitable types of polymer fibers and used in the sailboat building.

1.7.2 POLYMER COMPOSITES IN THE MEDICAL DEVICE INDUSTRY

Polymers and their composites in thermoplastics and thermosets are widely used in medical devices, replacing metals and glasses. The main reason for this is the versatility of polymers, which enables different applications by designing and producing a wide variety of products. The ability to make polymer blends, incorporating various performance-enhancing additional agents, contributes to the versatility of polymers.

Below are some examples of such uses:

- PMCs are used in instruments such as scanners, C-scanners, X-ray couches, mammographic plates, tablets, surgical target-insulating tools, and wheelchairs.
- Nanocomposite polymer matrix that contains carbon nanotubes or nanotubes with TiO_2 reduces bonehealing time by acting as a "scaffold," which guides bone replacement growth.

- The possible use of nanocomposites is being investigated in diagnostics and therapy. For example, the combination of magnetic nanoparticles with fluorescent nanoparticles in each of these magnetic and fluorescent particles seems to facilitate the tumor's visibility during the preoperative magnetic resonance imaging test. It may also help the operator better view the tumor.

1.7.2.1   Benefits Provided by Polymer Composites

Polymers and composites of polymer matrix have helped improve the quality of healthcare and save many lives by offering the following benefits:

- Disposable products not requiring sterilization
  - Polymers can produce cheap, disposable instruments such as syringes, catheters, and surgical gloves.
  - To ensure that the syringes were correctly sterilized after each use before disposable plastic needles were made available, as the patient was at risk of infection if the syringes were not correctly sterilized after use.
- Improve safety by making tamper-free packaging, such as tamper-resistant caps, available on medical packaging.
- Providing economical solutions for numerous medical devices.
- Improved comfort as people usually have more ease in contact with polymeric surfaces than contact with metal surfaces.
- The lightweight of metal and glass increases the ease of handling.
- Higher biocompatibility, important for applications such as implants.

1.7.3   Polymers and Polymer-based Composites in Different Applications: The Sporting Goods Industry

Sports goods: composites with polymer matrix find numerous uses in sports goods. A few examples are listed below.

- Because of their lightweight, high strength, many degrees of free construction, easy processing, and forming features, fiber-reinforced composites can be used as building materials in high-performance sports gear. Sports equipment, such as skis, baseball bats, golf clubs, tennis racks, and bike frames, are examples of such equipment that utilize composites.
- Some more examples of technical textiles in sports gear are parachutes, balloon fabrics, paraglide fabrics, and sailcloth.
- Polymers for protection of impacts and shocks, handgrips for motorcycles, muscle rollers, exercise mates, floating mallard decoys, and surfboard components are used in various gadgets padding or insert in motorcycle helmets.
- Protects against dangers posed by sports equipment, especially excessive humidity, promotes bacterial and fungal growth, biologically resistant or reactive polymers, and composite products.

Footwear: With the help of PMCs, it is possible to improve the performance and comfort of shoes and shoe's interior and exterior durability. Further, the types of inconvenience in conventional shoe textiles such as odor, bacteria, and fungi may be counter acted by biologically resistant or responsive composites. In high-quality footwear, synthetic (artificial) leather produced using polyurethane formulations is often used as natural leather alternatives. For the production of footwear, which can be used for extended periods in challenging conditions, the optimal use of PMCs is essential.

Performance footwear: The optimum use of polymers and PMCs are essential to producing high-quality, comfortable footwear that can be used for long periods despite stringent requirements during use. Polymers and PMCs are therefore widely utilized in enhancing footwear performance. Some examples are as follows:

- Among building performance, footwear materials are technical textiles of various designs, chosen mainly because of their technical and performance characteristics. Cotton, wool, nylon, polyester, polypropylene, rayon, spandex, and many other fiber types are among those used in textiles. Textiles made of these various fibers and their mixes cover a broad spectrum of characteristics and performances and are used in multiple performance shoes.
- Synthetic (artificial) leather made from polyurethane is often used in high-performing footwear as an alternative to natural (mostly cow) leather. It is available in various designs at various quality and cost levels. It is often made of two layers of woven or non fabricated polyester fibers, one of them being a backrest.
- In tongues, collars, and uppers of shoes, open-cell polyurethane foams with different densities and thicknesses are applied. If the maximum amount of ventilation is required, open-cell reticulated foams are used.
- Midsoles are produced from cell-closed foaming materials such as copolymer, polyethylene, styrene-butadiene, polyurethane, natural rubber, and polychloroprene. Midsoles are made from closed-cell foams.
- There are other advantages of polymers and composites used for performance footwear by using additives. The typical drawbacks of traditional shoe textiles such as odor, bacteria, and fungi are used to counter bio-resistant or reactive composites.

1.7.3.1 Building, Construction, and Civil Engineering Sectors

Examples of PMCs include the replacement, repair, upgrade, or reinforcement of a structural component made from a traditional fiber-reinforced material and the emerging composite panel technology for modular construction of buildings.

Two crucial mechanical properties are usually the rigidity (elastic modulus) and the strength of construction material. For instance, in many applications where such features must exceed specific threshold values for the material to be applied correctly, they cannot use a material of insufficiently high modulus and strength. Also crucial in some applications are the percent extension at the output (for materials that show an output point) and the ultimate elongation (the pressure where the material ruptures).

Furthermore, weight reduction becomes a primary and secondary selection considering candidate materials with sufficient rigidity and strength. The modulus/density ratios and strength/density ratios define "specific stiffness" and "specific strength." Polymers are of low density and weigh less than an equal amount of glass or metal in a given polymer volume. Many fiber-reinforced polymers (FRPs), which have adequate structural strength and stiffness for use, have higher specific rigidity and specific strength than metals because their density is far lower and, therefore, more suitable for this application. The superior resistance to corrosion in FRPs is often a further benefit.

In some applications, thermal properties are essential. The best example is thermal insulation materials in buildings with polymeric foams, which provide an exceptional thermal insulating effect due to their very low thermal conductivity. When a material is used as a protective cover, vapor barrier, sealant, or scaling compounds, its gas and vapor permeability is essential. When a material is used as an adhesive, it is essential to adhere to its adhesion and maintain adhesion durability in the application environment.

The materials must be resistant to fire. There are numerous standardized methods for evaluating the fire characteristics of construction materials. In specific applications, these features are more important than in others. For example, building code requires polymeric foams used in insulating the interior wall to be covered by a thermal barrier or other means of reducing fire risk. In contrast, plastic laminates that are used in countertops and kitchen cabinets are not required.

When materials are used for external constructions, it is essential to determine whether the material shows good weathering and aging resistance in the environment it is exposed to. Depending on this environment, it may be necessary to consider the effect of factors such as chemical exposure (such as acid rain exposure), heat, thermal shock, UV exposure, and high-energy exposure.

In the selection of materials, environmental sustainability is becoming more and more critical. Therefore, the expected environmental effects of various material choices, which comply with an application's requirements, have to be compared. Retrofitting a building with material replacement components is sometimes cost-prohibitive. Therefore, most building materials remain permanently in the buildings. Consequently, it is essential to remember that, because of various reasons for avoiding them that are either not first recognized, some of the building materials that were used extensively once are used not at all or very rarely used. Examples include asbestos, formaldehyde-containing thermal insulating foams, and poly (1-butene) pipes.

Finally, profit margins are essential factors in choosing substances that fulfill the performance standards among candidates.

Impellers, blades, housings, and covers:

- Windmill blades are among the numerous kinds of products that provide the necessary end-user performance with the high strength-to-weight ratio achievable through PMCs. Examples include blades made in an epoxy thermoset matrix polymer from carbon nanotubes or graphene nanoplatelets.
- Nanocomposites of this kind lead to electricity. The electricity of the nanofiller components depends on their distance. When strong winds cause the

blades to bend, the distance changes. Therefore, these structural components are also used as stress sensors to warn the windmill operator when the windmill should be shut down to protect against severe damage.

- Nanocomposites in the polymer matrix are also used as an impeller building material for many applications, including vacuum cleaners.
- The nanocomposites polymer matrix is also employed as building materials for several housing and coverings, such as power tool housing and padding hoods, as well as covering mobile electrical devices such as mobile telephones and pagers.

1.7.4 USE OF POLYMERS AND POLYMER-BASED COMPOSITES IN ENERGY-RELATED APPLICATIONS

Energy storage devices: In many energy storage devices, PMCs are used. Some examples are given below.

Modern phenomena include the emergence of the electric and electronics industries and the emergence of polymers as a new kind of material. The first electronic device was the electric relay and a remote, electricity-controlled switch invented in 1835. The first fully synthetic polymer was Bakelite, a thermosetting phenol formaldehyde resin developed in 1907. For a little over a century, the electrical and electronics and plastics industries have grown simultaneously.

In various electrical and electronic applications, thermoplastic and thermoset polymers and their composites are used to perform different functions.

This trend's primary reason is the remarkable polymer versatility, which makes it possible to design and fabricate a wide variety of products at an acceptable cost to fulfill different application requirements. The capacity to produce polymer mixtures, integrate numerous performance improvements, and produce PMCs by incorporating regenerative components (e.g., fibers, fillers, and particulate fillers) contribute to a greater versatility than is offered by individual polymers alone.

- Lithium-ion batteries have increased power output from anodes produced using the nanocomposite of silicon nanospheres and carbon nanoparticles. Other nanocomposite formulations have also been evaluated with favorable results following this initial study. It should be noted that these particular nanocomposites do not possess a polymer matrix, although they are mentioned here because of their importance.
- Paper may produce a conductive paper that may be soaked in an electrolyte to make flexible batteries. As the primary component of a paper is cellulose (a polymer), the conductive paper is a nanocomposite polymer matrix. This work is an essential step toward flexible (bendable) electronics in the emerging field.
- In the thin film condensers for computer chips, polymer matrix nanocomposites are used.

1.7.5 OIL AND GAS EXPLORATION, PRODUCTION, TRANSPORT, AND STORAGE

In many applications in the oil and gas industry, PMCs are used. Some examples are given below.

- Composites of FRPs are used as construction materials in oil and gas exploration and production platforms. The significant advantages of PMCs over metals for such applications include much lighter weight and more excellent corrosion resistance.
- In the last 20 years, oil and gas transportation and storage media, from high-pressure equipment and piping to oil storage tanks, have increased rapidly. Such composite repair systems are used as alternatives for replacing or repairing installed heavy metal sleeves and damaged steel pipeline components.

REFERENCES

Congress, U. S. "Advanced materials by design." Washington, DC: US Government Printing Office (1988).

Chawla, Krishan K. Composite materials: science and engineering. Springer Science & Business Media, 2012.

Mallick, Pankar K. Fiber-reinforced composites: materials, manufacturing, and design. CRC press, 2007.

Jaswal, Shipra, and Bharti Gaur. "New trends in vinyl ester resins." Reviews in Chemical Engineering 30.6 (2014): 567-581.

2 Natural Fiber-reinforced Polymer Composites

K. ArunPrasath
Kalasalingam Academy of Research and Education

P. Amuthakkannan
PSR Engineering College

CONTENTS

2.1 NATURAL FIBER AS REINFORCEMENT IN POLYMERS

Natural fibers are naturally available resources used as reinforcement materials for manufacturing a composite material. The main advantage of using these natural fibers as reinforcement material in composite materials is that they are abundantly available on Earth. Natural fibers are mainly available from vegetables and plants (lignin and cellulose), and other naturally available fibers are extracted through animal hair, skin, and mineral fiber [1]. Additionally, these natural fibers have good mechanical properties, but they are susceptible to moisture. Chemical treatment of natural fibers enhances the mechanical properties of the fiber [2].

The development of eco-friendly materials is essential nowadays, with many researchers focusing on the involvement of natural fiber as reinforcement material in the preparation of composite material. Agricultural waste such as husk, maize, cobs, and cotton silk are excellent residues and versatile alternatives for synthetic fibers in addition to plant and animal fibers [3]. The most utilized natural fibers to produce a composite material are jute, kenaf, bamboo, flax, ramie, coir, banana, sisal, cotton, pineapple, and oil palm; however, many natural fibers are not yet identified for proper utilization in composite manufacturing [4].

The manufacturing cost of natural fibers is very low and they also satisfy the industrial requirements by providing many potential benefits and more stability. The main

23

applications of natural fiber-based composites are in the fields of automobile, marine, aerospace, and some structural components [5]. The chemical treatment of natural fiber provides additional strength to the composite material by removing cellulose and hemicellulose content from the natural fiber.

Chemical treatment improves the mechanical properties of the fiber. Another naturally available component is natural resins, but the utilization of these resins is comparatively low because the processing and curing of these resins is difficult. However, with the current push for sustainability, focusing on manufacturing of green composites with naturally available materials helps to meet the industrial demands and environment [6].

In terms of mechanical properties, research has shown better performance of natural fiber-based composites which display good mechanical properties. For the better Mechanical property evaluation use ASTM standards and minimum of three trail averages for the tensile strength, flexural strength, and impact strength with the ASTM standards as D638, D790, and D256 [7]. Natural fibers provide a good interface for bonding between the polymer matrices and provide better load-bearing capacity to the composite. They not only have better mechanical properties but also improved electrical conductivity. The combination of jute fiber with polyvinyl alcohol (PVA)/graphene shows increased electrical conductivity because the interconnected crisscross channels present in the graphene sheet tightly bond with the jute fiber [8].

Natural fiber reinforcements also show the better wear characteristics when the sliding of frictional force increases. Natural fiber reinforcement composites possess resistance against further sliding which increases the frictional coefficient of the composite, which is responsible for providing better wear resistance. The wear path of natural fiber composite is always parallel to the high coefficient of friction which decreases the adhesive wear in the composite [9].

The machining characteristics of natural fiber-reinforced composite show improvement in speed, feed, and tool geometry. The cutting behavior of natural fiber composite differs from that of the synthetic fiber composite because of the cellulosic structure of natural fibers. However, the results may vary with changes in fundamental characteristics such as orientation of fiber, cutting direction, chip shape, along with other morphological behaviors. Comparatively better results are observed with the natural fiber as reinforcement in the composite [10].

Cellulosic natural fibers provide promising thermal stability with good crystalline property when they are utilized as effective reinforcement for composite materials. Natural fibers offer good aspect ratio and stiffness when combined with the polymer matrices. The large hollow structure of natural fibers has the ability to absorb more sound effectively, with the lumen structure of the natural fiber providing good sound blocking than other sound absorbers [10, 11].

Dynamic mechanical analysis (DMA) of natural fiber-reinforced polymer composite show effective stress transfer by increasing the adhesive bonding between the matrices. The addition of natural fiber in the composite provides high resistance toward the molecular movement for improvement in the damping property of the composite [11].

The low velocity impact response of natural fiber-reinforced composite shows concrete results. When the number of layers of natural fiber content increases in the composite, they act as a barrier during low velocity events and provide small perforation on the surface of the composite. In comparison, the low velocity results of natural fiber-reinforced composite show improved results with polymeric matrices [12].

2.2   SYNTHETIC FIBER AS REINFORCEMENT IN POLYMERS

Synthetic fibers are manmade fibers made by chemical synthesis processes; these material fibers are extracted from chemical or petrochemicals. Then they are polymerized for improved physical and chemical properties. Although natural fiber composite is one of the better alternatives for synthetic fibers, they are not fit for moisture resistance, fire resistance, and some surface modification-related applications. The addition of synthetic fiber with natural fiber enhances the moisture and fire resistance of the composite. Adding synthetic fiber in the composite helps to delay the failure initiation of the composite during load application in mechanical conditions.

The combination of natural and synthetic fiber composite produces better mechanical properties; however, natural fibers are a better choice. The addition of synthetic fiber increases the matrix strength and reduces the shear stress due to the application of external load. Equal contribution of both natural and synthetic fiber increases the aspect ratio of the composite [13].

Liu et al. [14] reported that the main objective of adding synthetic fiber with the reinforcement natural fiber is to reduce the damages on the surfaces due to sudden loading conditions. Experimental and theoretical studies of reinforced concrete with the combination of steel and synthetic fiber composite show enhanced results with different dosages of synthetic fiber with steel concrete.

Crack movement is determined by the strain rate of the composite; the tension is equal to the cracking strain of the reinforced fiber and steel concrete. While measuring the crack dimension of the composite, 80% of synthetic fiber dosage affects the volume of the fiber and resists the post cracking stiffness with enhanced results. Massive amounts of natural fiber available in the world can be utilized for various applications. The utilization of waste agricultural products to convert useful products is a novel technique today. Addition of eco-friendly synthetic fiber enhances properties and their behavior for lightweight applications [15].

Currently, concrete is one of the structural materials widely used for construction. High-strengthened concrete mixed with steel and synthetic fiber reduces the weight and increases strength. However, when concrete is very strong, it becomes brittle in the absence of any additives even though strength is improved. Synthetic fiber and concrete mixture improves the strength of the composite with increased energy absorbing capability by reducing the surface damage.

Fallah-Valukolaee et al. [16] investigated the stress–strain behavior of various combinations of macropolymer/polypropylene fibers with hybridization in the presence of nanosilica. Addition of 1% of polypropylene improved compressive strength compared to the other two combinations. This result demonstrates the application of concrete without steel and other normal aggregates utilized in construction. Polypropylene-based concrete demonstrates better resistance to microcracks and increases the compressive strength than plain concrete. This shows that synthetic fiber-based concrete structures are better alternative than steel-based plain composite with more strength.

Utilization of synthetic fiber for various applications is a new trend in composite materials; however, lifecycle analysis of the hybrid natural and synthetic fiber composite is lacking. Moreover, natural fibers are composed of lignin, cellulose, and hydroxyl ion which degrade over time. To reduce this degradation, they undergo chemical treatments for property enhancement and for removal of waste products from natural fibers. Synthetic fibers have some advantages over natural fibers.

Deb et al. [17] evaluated the mechanical properties of cement and mortar reinforcement with jute fiber and polypropylene fiber with their lifecycle cost analysis. The individual fiber composite's mechanical property was low compared to the hybrid composite. The improved mechanical properties are noted for the combination with higher amounts of polypropylene fiber. The lifecycle cost analysis shows that cementitious polypropylene fiber reinforcement composite performs better than the jute reinforcement composite.

The moduli of elasticity of different composites is one of the factors to consider when evaluating the different mechanical and lifecycle cost analysis of any of the composite combinations. The lowest value shows the reduced property of the composite; however, 50% mixtures of both natural and synthetic fiber hybrid composite show improved results compared to other combinations. While comparing the individual fiber composite, hybrid combinations of the fiber improves the mechanical properties of the synthetic fiber composite and results with good ductile strength and shows better adhesion between the reinforcement and matrices.

The mechanical behavior of fiber composites is influenced by parameters such as fiber length, volume fraction, orientation, fiber and matrix adhesion, and interfacial bonding between the surfaces. Better bonding between the reinforcement and matrices provides better strength for the composite. The flexural strength analysis of bamboo and epoxy composite shows lower values at lower fiber content. In case of insufficient matrix to transfer the load, the treated bamboo fiber shows better mechanical properties compared to untreated fibers. Excess amount of fiber content in the composite deteriorates the mechanical properties. This shows that improper bonding in composites lowers the mechanical behavior. In the case of treated and untreated fiber reinforcements, the treated natural fiber with synthetic polymer results in better mechanical performances [18].

Yoo et al. [19] investigated the effect of synthetic fiber content, matrix strength, and heating spalling resistance of polypropylene and nylon reinforcement with cementitious composites. The results showed that addition of polypropylene and nylon fibers enhances the spalling resistance of the composite. Higher amount of the matrix's strength improves the compressive strength of the composite. However, decreased toughness with enhanced flexural resistance is noted for higher amount of reinforcement with the matrices.

Although the fire resistance of synthetic fiber is better than the natural fiber composite, the combination of natural fiber with synthetic fiber shows good results. Low permeability increases the strength of the composite at around 80 to 100 MPa with the permeability value ranging between 10^{-15} and 10^{-16} m/s. The average heat withstanding ability for the steel bar is below 649°C for the heating period of 3 hours. The heating resistance of polymeric-based fibers such as polypropylene and polyvinyl alcohol shows better fire resistance than normal metals at elevated temperatures.

Composite materials attract attention as replacements for conventional materials like aluminum, steel, and copper. Such applications lead to the introduction of synthetic polymers (glass, carbon, aramid, etc.) for wear applications. The sliding speed, applied load, and sliding distance are considered as the recommended parameters to study the wear characteristics of the materials. Polymeric materials are more brittle and exhibit good wear resistance than metals. Kevlar reinforcement with epoxy matrix shows improved ductility and adhesion between the fiber and matrix. The chopped E-glass combined with Al_2O_3 shows increased young's modulus and tensile strength of the composite [20].

The chemical and corrosion resistance of polymer is higher than metals, and some of the synthetic polymers are widely available at less cost. The durability of synthetic fibers is better than naturally available fibers. The mechanical behavior of these polypropylene fibers shows good results with resistance toward deformation with improved crack, ductility, compressive, and bending. The increased synthetic fiber dosage in the composite improves fracture and residual tensile strength.

Minhaj et al. [21] reported that the elastic modulus of synthetic fiber dosage is affected by the orientation of the fibers. Synthetic fiber placed parallel to the direction of the load application displays reduced elastic modulus. Greater energy absorption capability of synthetic fiber under stress shows increase in the fiber dosage and reduced average toughness of the composite. The stress–strain curve shows some good peaks on energy dissipation and ultimate strain for synthetic fiber composite.

In general, synthetic fibers have good properties than natural fiber composite as well as good strength and stiffness. Carbon fiber is one of the synthetic fibers used for multifunctional applications. They have enhanced thermal and electrical conductivity, and the main advantages of carbon reinforcement composite include low weight, higher stiffness, higher strength, and chemical resistance. However, the cost of purchasing and processing of carbon fiber is higher compared to other synthetic fiber reinforcement composite.

Short carbon fibers are exclusively used for automobile, marine, and other structural applications. Owing to their strength-to-weight ratios with attractive mechanical properties, polymer matrix composite produces good results. By varying the weight percentage of the carbon fiber content in the composite, better tensile strength is noted at different load transfers. Whenever the carbon fiber makes an interfacial bond with the polypropylene matrix it produces a better result. Moreover, good elongation break is also noted on the elastic regions with ultimate strength of the polypropylene matrix at higher loads [22].

2.3 PARTICULATE AS REINFORCEMENT IN POLYMERS

Composite materials are filled with particulates in the form of powders which provide additional strength for the composite to meet the demand. The particulates are categorized as natural and manmade. Natural particulates are manufactured from fibers, plant stems, seeds, and wood. Synthetic particulates are made from metals and

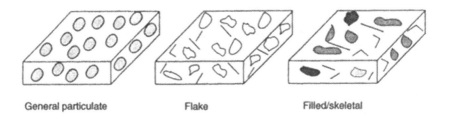

General particulate   Flake   Filled/skeletal

FIGURE 2.1   Some forms of particulate composites.

manmade fibers such as aluminum, boron carbide, and tungsten. Metal particulates and all types of crashed fibers are used to produced particulate-based composites for various applications. Size of the particles are less than 0.25 μ for particulates, which may vary according to the product requirement and applications. Particulates are classified into three categories, namely, general, flake, and filled. Flake particulates with large area of thickness and skeletal or filled particulates are considered as secondary matrices to provide sufficient strength for the composite. Figure 2.1 illustrates some forms of particulate composites.

Wood flour waste is one of the major byproducts of plywood, furniture, craft industries, building components. These waste wooden particles are used as landfills in industries and converted in the form of energy by approximately 40%–50% without affecting the environment and ecology. Epoxy and polyethylene resins are commonly used matrix materials which produce better bonding and enhance mechanical properties with wood-based polymer composites [23].

Oladele et al. [24] investigated the snail shell particulate and waste plastic (polypropylene) for automobile applications. Thermoplastic materials are lightweight materials which are perfectly suitable as automotive parts. Effective stress transfer is 12–15 wt.% of snail shell particulate and polypropylene matrix. This filled particulate is the key strengthening mechanism for tensile strength owing to the presence of calcium carbonate in the snail shell which improves the composite's resistance toward deformation.

Sharma et al. [25] discussed the fracture toughness of the glass-filled epoxy composite under impact loading conditions. The varying weight percentage and effect of filler shape was investigated using artificial neural network. Artificial neural network is one of the exciting tools to study a wide range of engineering and mathematical problems. Experimental models with 91% of glass-filled epoxy composite show enhanced dynamic modulus and volume fraction. During sudden impact loading conditions, the rod-shaped, glass-filled epoxy composite exhibits better toughness resistance toward crack initiation. This observation shows that the higher aspect ratio and stiffness between the interfaces are responsible for enhanced dynamic mechanical properties.

The recommended particulates to mix with polymer composite are talc, red mud, clay, fly ash, alumina, and silica. These particulate matrices improve the mechanical properties of the composite by enhancing the strength and stiffness. Micro Al_2O_3 is used as an additional matrix with woven glass and epoxy composite. Improved wear and mechanical properties are noted, and the mixture of inorganic particulates produces a homogenous structure and supports the composite at varying dynamic loading conditions [26].

Srinivasan et al. [27] investigated the mechanical properties of banana fiber/epoxy composite filled with calcium carbonate particulate. Addition of calcium carbonate as matrix material in the composite combination results in unique characteristics such as specific gravity, low cost, good modulus of elasticity, and good mechanical properties. By varying the weight percentage of calcium carbonate from 3 to 6 wt.%, the composite is produced via compression molding technique. The composite filled with 6 wt.% of calcium carbonate shows better mechanical properties because calcium carbonate particulates are completely dissolved in the matrices and fill the voids in the composite. Hence, stress flow inside the matrices is smooth and provides better strength to the composite.

The chemically treated natural fiber possesses better mechanical property than untreated natural fibers. However, the presence of lignin and some pigments affects the behavior of the fiber at different loading conditions. Animal natural fiber and hairs are chemically treated for enhancing performance. Naturally available animal fibers have better strength. Moreover, fibers derived from sea-based living organisms exhibit good strength either with treatment or without chemical treatment. With the same results, chicken feather-reinforced composite produces better mechanical properties after chemical treatment.

The chemically treated antennas of prawn reinforcement with polyester matrix filled with waste plastics such as hydrogen peroxide, potassium hydroxide, and sodium hydroxide has been demonstrated. The addition of 10 wt.% of treated fiber with 15 wt.% of waste plastic combinations shows better tensile strength. For flexural strength, better results are noted for 10 wt.% of treated fiber with 5 wt.% of waste plastic, and the same combination displays better impact characteristics. Adding particulates in the composite increases the strength of the composite and shows better mechanical properties with chemically treated natural fiber composite [28].

The waste recycling of industrial leather-based composite is prepared with a combination of epoxy matrix and high-density polyethylene. Waste leather is crushed into particles and added to the composite by varying the weight percentage. The compression and fracture toughness of the composite is measured by adding these leather particles, and 8% and 65% improved results are noted than the pure form of epoxy/high-density polyethylene composite [29].

To understand the failure mechanism of the composite, fractography technique is used and the change in the failure mode is observed. Normally, failure strength of pure composite without any particulate mixtures is very low compared to the composite filled with particulate. This occurs when the bonding between the interface is good. Naturally available animal-based particulate possesses better mechanical properties when blended with polymer composite. The efficacy of particulate dispersing in the composite is more and they are effectively utilized as a successful composite product.

Jeyapragash et al. [30] investigated the mechanical properties of natural fiber and particulate-reinforced epoxy composite. The common bast natural fiber includes a wood core on the surface of the stem. The inside of the wood core is filled with large amounts of fiber bundles, which have more strength and provide support to withstand applications at external loads. The hybridization of coir natural fiber filled with egg shell and rice husk particulate shows enhanced mechanical results than the unfilled composite. Improved bonding and correct form of blending mechanism of particulate in the matrices are also observed.

Fly ash-based particulate fillers mixed with polyurethane and carbon fiber show improved mechanical properties. Varying the weight percentage of fly ash content in the composite displays different properties. In that, 15% of fly ash particulate results in better mechanical performance compared to other combinations. The perfect distribution of particulates in the matrix shows this observation on the composite's mechanical properties. The main advantages of sandwich composite laminates on fiber-reinforced composite are high performance and uniform distribution of particulates for enhanced performance [31].

Inclusion of particulates in the polymer matrices with perfect distribution shows enhanced results. This also confirms the better elastic property with less amount of degradation and better agglomeration. The cell geometry of particulates is analyzed using morphology analysis technique. Better distribution and constrained pattern are essential for better mechanical properties of fiber-reinforced composite.

Prasath et al. [32] investigated the low velocity impact mechanical behavior of basalt powder and woven flax fiber polyester composite. Basalt powder particulates contain rich volcanic minerals which influence the impact property of the composite when they are combined with the natural flax fiber. The composite combination with 10 g of basalt fiber and 10 layers of flax fiber provides better energy absorption capability than other combinations. The particulate basalt powder supports the composite with some residual energy when impact events occur. Overall, the addition of basalt powder particulate produces better results with flax fiber polyester-based composite.

2.4   FILLERS AS REINFORCEMENT IN POLYMERS

In the earlier years of composite material, significant research was done using synthetic and natural fibers as reinforcing material. These natural and synthetic materials provide excellent mechanical properties and related applications. The interfacial adhesion of filler materials with reinforcement and matrices produces good mechanical and tribological properties [33].

Chaudhary et al. [33] studied the effect of graphite filler on palm/epoxy composite with the addition of 3 wt.% of graphite as filler material, which increased the wear performance of the composite. However, higher graphite content affects the wear performance of the composite. The blending nature of the filler with the fiber is one of the influencing factors which decides the property of the filler-supported composite. Adequate adhesion between the filler and the reinforcing material produces better bonding without voids which is confirmed through scanning electron microscopy (SEM).

In fiber reinforcement composites, it is challenging to develop a composite with cellulose because of its poor compact ability and hydrophobicity. To overcome this issue, cellulosic filler materials are rendered with hydrocarboxyl content to improve the quality of the filler. The surface modification of the filler produces optimum results when combined with polypropylene matrices. The maximum flexural modulus is noted for 4 wt.% of cellulose, and the adverse impact produces excellent lubrication on the samples during load application. Systematic reference with the combined chemical and mechanical properties improves the introduction of filler-based industrial products [34].

The specific qualities of advanced polymer composite lead to the incorporation of fillers in the composite materials. This research highlights the use nanofillers toward improved thermal characteristics and low-density essential components. Adding nanofillers to the composite reinforcement enhances mechanical properties. Nanofillers also act as a supporting layer to prevent the composite from damages and synergy and provide good electrical conductivity. They also offer higher performance with hybrid thermoplastic and thermoset polymer matrices.

Hemath et al. [35] reported that the addition of carbon nanofillers in fuel cells increases the thermal conductivity inside the fuel cells. The thermal conductivity of the polymers lies between 0.1 to 0.3 W/mk before the addition of nanofillers in the composite. Bonding strength and flexible polymer movement in the composite were noted after the addition of 5 wt.% of nanofillers in the composites. The essential stress relaxation and energy dissipation of polymer molecules is enhanced after the addition of nanofillers which provides better electrical conductivity to polymer composites.

Owing to the good mechanical properties, quality, durability and thermal stability, the coir-based fillers are widely used. The composite is prepared with the epoxy matrix and coir as a filler material, and its wear performance is measured. The wear load varies from 10 to 40 N, and increasing coir content (2.5%) in the epoxy matrix provides better results. The greater filler reinforcement provides steady volume loss in both dry and wet conditions. The SEM observation results with microcracks and wear fragments represent the perfect bonding nature of the filler with the polymer matrices [36].

The addition of recycled rubber elements in metal powders increases the mechanical performance of the composite. Especially, addition of aluminum powder results in property enhancement compared to other metal-filled rubber composite. The polymer matrix composite is prepared using acrylonitrile butadiene styrene (ABS) as matrix and fine particles of aluminum filler (250–800 μm) as reinforcement. The mechanical properties of these composites are tested with various proportions of aluminum powder in ABS and show better mechanical properties. With the help of stress and strain curve, the tensile strength of the composite is examined. The composite with higher concentration of aluminum shows better mechanical properties. The presence of aluminum metallic fillers controls the movement of elastic rubber to develop an interfacial adhesion and dispersion between the filler and matrices.

The SEM morphology indicates the homogenous mixture of fillers in the rubber based on the cracks formed on the surface of the sample, representing the significant mixture of organic and inorganic interfaces in the composite [37].

Biochar is one of the effective non-renewable filler materials which is extracted through naturally available organic materials. It is made by burning biomass under a controlled pyrolysis process which contains more amount of carbon substances. Addition of various physical properties of biochar with the polymer matrices produces good interfacial bonding and strength. This biochar is prepared from wood dust, corn dust, and other organic matters through a controlled process.

Matykiewicz [38] investigated the DMA of biochar and epoxy matrix composite. Viscoelastic properties of biochar and epoxy matrix composite results with continuous changes in the storage modulus and damping factor values were seen. The stiffness of the composite is increased by adding 10 wt.% of biochar with the temperature value greater than 50°C, which shows a glassy region with significant drop in the damping factor value and the higher degree of energy dissipation. The introduction of biochar filler with epoxy matrix improves the thermal stability of the composite.

Dielectric properties of the materials have several applications in semi-conductors and energy storage devices. In the automobile industry, hybrid vehicles are manufactured using polymer composites. Usage of ceramic powder during manufacturing of polymer composites produces better dielectric properties. However, the ceramic materials are brittle in nature, large quantities compromise the strength and integrity of the composite. Coir pith added with 50 wt% of ceramic filler shows improved dielectric property. The addition of filler ceramic shows improved results with better conduction of electricity. Addition of biochar in the composite increases the fire retardancy, flexibility, and toughness of the material [39].

Natural seaweed powder (brown algae) with particle size ranging about 30 μm is used as filler material. It is effectively mixed with carbon fiber reinforced with epoxy for composite development. The tribological studies and tensile modulus of the composite is evaluated. With increases in the dry sliding velocity of the composite, the wear rate also increases. However, the minimum surface wear is obtained for the composite which has 10 wt% of filler powder. The development of low stress on the surface of the composite resisting the material's plastic deformation with less amount of wear is noted for 10 wt% of filler-filled composite. In the case of mechanical properties, the same combination shows improved flexural strength. The addition of naturally available seaweed filler is a newer effort for developing composite materials which is useful in tribological and tensile applications [40].

A novel lithophilic polyacenequinone polymer fillers are more flexible in design and shows very good mechanical property. Fei et al. [41] investigated the energy conversion and dissipation properties of lithophilic gel polyacenequinone polymer fillers. Compared to traditional lithium ion batteries, these polymer-based electrodes produce better metallic compounds and energy conversions. Based on rapid discharging cycle and low columbic efficiency, the usage of this organic polymer fillers offered better conductivity and polarization. The design methodologies and molecular structure of the polyacenequinone polymer filler act as a replacement for metal lithium anodes. In the future, polymer will be the only replacement for metals and their constraints.

The aggregation of filler with nanocomposites and microcomposites increases the electrical conductivity of the polymer matrices. The addition of graphite particles with polyvinyl chloride and polypropylene matrices in the form of nanosheets and nanotubes for the composite preparation to achieve better results. Addition of graphite filler with these matrices improves electrical properties. Polymer and filler particles are embedded together to form an isentropic matrix for good conductivity. However, the effective conductivity of the filler material increases the volumetric fraction of the compound and acts as a bridge between the filler and the polymer matrices. Two models for testing conductivity produced from polyvinyl chloride/graphite filler and polypropylene/graphite filler have been compared. In the form of the sheets and tubes, both combinations show improved results with maximum polarization can be achieved with better conductivity [42].

2.5 APPLICATIONS

Natural fiber-based hybrid composites have superior properties than synthetic fiber composites. Natural fiber products have more applications which are specifically utilized for various products. Better replacement for the metal–polymer matrix composites is one of the alternatives. Lower weight with superior strength produces a unique property which is used for making aircraft wings in aircraft manufacturing industries. In shipbuilding industries, the hull of the ships and some of the interior frames are manufactured with fiber-reinforced composites.

In automobile industries, most of the essential parts are now built with polymer matrix composite. Specific parts like frames, dashboards, doorsteps, windows, and doors are manufactured with natural fiber composite. Polymer matrix materials have some commercial applications such as manufacturing of toys, fiber concrete structures, swimming pool panels, racing car bodies, shower stalls, bathtubs, and storage tanks, which are produced through natural fiber composite. James et al. [43] discussed the utilization of natural fiber for automobile application for making doors and some essential products.

Especially in the European market, the various segments of automotive interior and exterior parts are filled with natural fibers with thermoset and thermoplastic polymers. Natural fibers such as kenaf, flax, hemp, sisal, and jute are predominantly used to produce automotive parts such as door panels, seat backs, headliners, package trays, and dashboards. Glass fibers are used in the automobile interior and exterior parts production with good mechanical properties and enhanced structural durability. Natural fiber composites will continue to expand their role in automotive applications with some challenges in moisture stability, fiber polymer interface compatibility and consistent, the technical challenges are rectified which is more useful to utilize the availability of natural fiber sources in more products by automotive manufacturers in the world.

Tailor-made lightweight natural fiber products are highly recommended for structural parts in air planes and train components. The biological nature and eco-friendly behavior of natural fibers can be incorporated for more applications. Cellulose fibers can provide better stability toward load application. The presence of semi-crystalline cellulose microfibrils in the walls of the fiber embedded in the matrices is responsible for enhanced strength in the composite Polyaniline, carboxy methyl cellulose, hydroxypropyl cellulose, and acetoxy propyl cellulose have good conductive properties which are used in sensors. Polypyrrole mixed with wood dust increases the conductive property of the composite which is seen in electro-optical cells to control the switch timings. Synthetic polymers like polystyrene is one of the effective derivatives which preserves the drugs from chemical reactions, bio-composites like hydroxypropyl methyl cellulose is used to prepare substitutes for drugs [44].

Faris et al. [45] investigated about the available natural resources and waste for sustainable utilization as some industrial products. They selected date palm fiber (DPM) and investigated its comprehensive applications toward industry and automobile engineering. The design of low-cost sustainable material production has been resolved with the help of natural fibers. Because DPM has better vibration absorbing capability, in automotive industry, components like spring and shock absorbers are manufactured through this fiber as raw material. The door panels of Mercedes-Benz E-class is made through flax/sisal fiber mat implanted with epoxy resin. Moreover, around 60% of automotive doors are manufactured with the natural fiber composite. E-glass is one of the suitable materials in composite technology with several industrial applications due to its remarkable strength, stiffness, and elongation at failure.

Organic materials are abundantly available on Earth with very good desired properties, Industrial and agricultural residues are contributed more in the property increment. Vaisanen et al. [46] studied the probable application of organic waste and its residuals with natural fiber composite. In this some of the waste management methods like composting and land filling were also added for better application. The effective utilization of possible waste recycled into products will help clean the environment. Due to the low density and cost with relatively high strength, natural fibers are used to produce some modified plastics. These natural fibers are added with additives to produce tailor-made complex industrial structures. E-glass, Kevlar and carbon fibers utilization is completely replaced by natural fibers with some low-density polyethylene for some useful industrial applications.

Sanjay et al. [47] discussed the commercial and engineering applications of natural fiber-reinforced polymer composites. Areca plant is one of the naturally available plant mostly seen in southwest India. These areca fibers are widely used in many applications, especially to produce medicine, paint, and chocolate. Another well known and oldest fruit is banana which is rich in vitamins C1, B1, B2, B6, E and minerals. It has various medicinal and mechanical applications because of specific properties like low density, low cost, high mechanical strength, stiffness and high growth rate. Several automotive parts are produced mainly in the adoption of polyester and polypropylene with fibers like flax, hemp, and sisal.

German automotive leaders like Mercedes, BMW, Audi, and Volkswagen are already on the path toward the production of products from natural fibers for interior and exteriors of their vehicles. The first fiber-based products was incorporated in S-Class Mercedes-Benz as an inner door panel in 1999. It was made through 5% Baypreg F semi-rigid elastomer from Bayer and 65% blend of flax, hemp, and sisal. Plastic and wood fiber composites are mostly used to produce large decks, docks, window frames, and some molded components.

Lignocellulosic natural fibers are added with polymeric matrices for mechanical and complex applications. Natural fibers are processed with thermoplastic matrices like polyethylene, polypropylene, and polystyrene to produce geometrically and rheologically stable products. In general, natural fiber composite has the ability to produce pseudoplastic behavior, which influences the orientation and load distribution capability of the composite to produce highly stabilized morphological and mechanical properties. Thermosetting matrices like phenolic, unsaturated polyester, phenol-formaldehyde, novolac-type phenol-formaldehyde, and epoxy resins produce composite materials that are added to natural fibers.

This type of combination (natural fiber + thermoset resin) has a unique property of withstanding the heavy impact load and improved fatigue behavior. Natural fibers like flax, jute, sisal, kenaf, hemp, cotton, and coir combined with thermosetting matrices produced load withstanding property. Increasing interest on natural fiber composite makes the productivity and application of these products a support for today's technological development and the associated applications [48].

In the past decades, researchers have been focusing on the utilization of natural plant fibers such as jute, sisal, coir, banana, hemp, kenaf, flax, and some other cellulosic fibers hybrid with synthetic fibers like glass, carbon, Kevlar, aramid, and others. The combined effect of both natural and synthetic fiber produces a good mechanical property, and the presence of natural fiber increases the wear performance of the composite. The hybrid effect of composite combination which has maximum contribution of natural fibers such as cotton, sisal fiber, and flax fibers are mainly used to produce furniture, tennis rackets, fences, bicycle frames, automotive door panels, and roofing sheets.

Electrical appliances, textile industry, mobile cases, insulation bags, mirror casings, brooms, and packing materials are some of the applications of natural fibers such as hemp, kenaf, coir, jute, and ramie fiber. These types of natural fibers have a greater potential to produce lightweight engineering and industrial applications [49].

Kumar et al. [50] investigated the challenges and opportunities available for natural fiber-reinforced polymer composites for industrial applications. The natural fibers are biodegradable, lightweight, cost-effective, and environment friendly. The effective utilization of natural fibers by focusing on automotive and furniture industries is discussed in this study. The commonly seen natural fibers like bamboo, cotton, flax, hemp, abaca, and coir display better properties for automotive and furniture manufacturing industries.

Kenaf and glass fiber-reinforced with epoxy hybrid composites are used to produce car bumper beams with more mechanical strength than metals. Jute fibers are one of the potential replacements for glass fibers which are environmentally stable. Jute fibers are used to manufacture front bonnet of vehicles. Polylactic acid and pineapple/cassava flour composite is very useful for automotive components. Abaca and jute fibers have better polymeric properties to produce mudguard and engine cover of automobiles. Borassus/high-density polyethylene composites are utilized as packaging material as well as in structural applications. The hybrid composites have a synergetic effect on natural fibers focusing on future studies and its applications.

Nayak et al. [51] investigated the potential utilization of natural fiber composite for ballistic applications. The use of ballistic jackets is to safeguard humans from sudden shock and impact loads. Multilayered ballistic armor system (MBAS) is one of the protection systems which consist of ceramic in the front and are successively filled with lightweight natural fiber composite layers for protection. Natural fiber is one of the key materials which has reasonable strength to absorb some immediate load. Similarly, bamboo fiber composite shows lesser performance about 18% less effective than aramid fiber composite.

With respect to price, weight, and environment eco-friendly, bamboo fiber is more efficient. Nettle natural fibers show better energy absorption and velocity ranges between 623.97 m/s and 837 m/s with the target thickness of 15 and 25 mm. This result shows there are several requirements for natural fibers reinforcement in ballistic applications.

Some of the new natural fibers like fique plant reinforced with polyester matrix show improved viscoelastic and thermal behavior. Moreover, it has very good hardness with the indentation depth of 15 mm in comparison to 23 mm, and the cost of these material preparation is very less (13 times) compared to polyester/Kevlar composite which is predominantly used for armor ballistic applications. Ramasubbu and Madasamy [52] studied the natural fiber characterization and its suitability for automotive applications. According to the international protocol, automotive manufacturers are looking for newer materials to reduce pollution and fuel consumption. In addition, these materials support recycle with more lifecycle assessments.

In the aeronautical field, technology has developed to utilize more composite materials-based products in the manufacturing of cockpits, wings, and winglet are major components produced from carbon and aramid-reinforced epoxy composite. In general, the price of the carbon and aramid fibers are more compared to natural fibers. However, the strength of natural fibers like jute, flax, and basalt is reasonable compared to synthetic fibers. So, the modification for these synthetic fibers is natural fibers which has specific application with more commercial inputs. The remarkable tensile strength and behavior of natural fiber in varying loading conditions are suitable for making some components in aircrafts and automobile engineering. Moreover, the availability and ease of processing natural fibers have a wide requirement in the replacement of synthetic materials, especially in the manufacturing of structural composites and related applications. Table 2.1 shows the some of the applications of fibers (natural and synthetic) with polymeric matrices.

TABLE 2.4
Specific Applications of Fiber-reinforced Polymeric Matrices

S. No	Name of the Fiber/Matrices	Specific Application	References
1	Glass fiber/epoxy matrices	Automotive parts	[43]
	Cotton fiber	Textile and yarn industries	
2	Polypropylene/wood flour	Electrooptic cell panels	[44]
3	Date palm fiber	Damping absorber, spring	[45]
	Sisal fiber	Construction industry	
4	Kevlar/carbon/LDPE	Manufacturing of industrial plastics	[46]
	Hemp fiber	Furniture and paper industry	
5	Areca plant/bayperg F elastomers	Medicine	[47]
	Kenaf fiber	Mobile cases, and insulation panels	
6	Jute, coir, flax/thermosetting matrix	Automotive decks, doors, and decorative items	[48]
7	Natural fiber + synthetic fiber hybrids	Furnitures, tennis rackets, textile, brushes, brooms, roofing sheets, and fencing nets	[49]
8	Kenaf, glass and Borassus/HDPE	Automobile front bonnet and packaging industries	[50]
9	Ceramic\bamboo fibers	Ballistic jackets, building panels, and chip boards	[51]
10	Kevalar/polyester	Automotive parts	[52]
	Coir fiber	Mirror casing and automobile seat cushions	

2.7 CONCLUSIONS

In recent years, fiber polymer composites have been chosen for various applications due to their availability, strength, durability, and good mechanical property. In this chapter, predominantly used natural fibers like flax, jute, cotton, coir, abaca, bamboo, kenaf, hemp, banana, pineapple, and many more fiber's reinforcements with polymer matrices and their property enhancements were discussed. Similarly, synthetic fibers like glass, basalt, Kevlar, aramid, carbon, and other fibers are discussed. Addition of particulates and fillers as reinforcement and some of the specific applications of these fiber-based polymer composite materials are also discussed. From the overall discussion, owing to the abundant availability, natural fibers have excellent mechanical, wear, and ballistic performance, which has wider application in the replacement of synthetic fibers and metals. However, natural fibers have some disadvantages like moisture absorption, low thermal stability, and degradation. Synthetic fibers are an alternative to natural fibers in some specific functions. Particulates and filler materials improve mechanical and dielectric property when they are mixed with natural fibers. Because of the size and chemically stable nature, these materials display improved properties. Application of replacing metals with fibers are more in various fields to produce automobile parts such as door panel, bonnet, interior, and exterior essentials of the vehicle are made with jute fiber/ epoxy matrix. In the aircraft industry, the airplane wings, cockpits, and winglet are made with jute, flax, and hemp fiber with polymeric matrices like polylactic acid and polypropylene. In general, the combination of natural fiber and polymeric matrices has larger requirements and benefits. In further extension of the research work, utilizing these natural fiber/polymer-based materials results in more eco-friendly products with different properties and for several applications.

REFERENCES

1. Navaneethakrishnan, G., Karthikeyan, T., Saravanan, S., Selvam, V., Parkunam, N., Sathishkumar, G., & Jayakrishnan, S. (2020). Structural analysis of natural fiber reinforced polymer matrix composite. *Materials Today: Proceedings, 21,* 7–9.
2. Suresh, S., Sudhakara, D., & Vinod, B. (2020). Investigation on industrial waste eco-friendly natural fiber-reinforced polymer composites. *Journal of Bio-and Tribo-Corrosion, 6*(2), 1–14.
3. Tavares, T. D., Antunes, J. C., Ferreira, F., & Felgueiras, H. P. (2020). Biofunctionalization of Natural Fiber-Reinforced Biocomposites for Biomedical Applications. *Biomolecules, 10*(1), 148.
4. Jeyapragash, R., Srinivasan, V., & Sathiyamurthy, S. (2020). Mechanical properties of natural fiber/particulate reinforced epoxy composites–A review of the literature. *Materials Today: Proceedings, 22,* 1223–1227.
5. Zaini, E. S., Azaman, M. D., Jamali, M. S., & Ismail, K. A. (2020). Synthesis and characterization of natural fiber reinforced polymer composites as core for honeycomb core structure: A review. *Journal of Sandwich Structures & Materials, 22*(3), 525–550.
6. Joseph, J., Munda, P. R., Kumar, M., Sidpara, A. M., & Paul, J. (2020). Sustainable conducting polymer composites: Study of mechanical and tribological properties of natural fiber reinforced PVA composites with carbon nanofillers. *Polymer-Plastics Technology and Materials, 59*(10), 1088–1099.

7. Chegdani, F., Takabi, B., El Mansori, M., Tai, B. L., & Bukkapatnam, S. T. (2020). Effect of flax fiber orientation on machining behavior and surface finish of natural fiber reinforced polymer composites. *Journal of Manufacturing Processes, 54,* 337–346.

8. Sumesh, K. R., Kanthavel, K., & Kavimani, V. (2020). Peanut oil cake-derived cellulose fiber: Extraction, application of mechanical and thermal properties in pineapple/flax natural fiber composites. *International Journal of Biological Macromolecules, 150,* 775–785.

9. Maran, M., Kumar, R., Senthamaraikannan, P., Saravanakumar, S. S., Nagarajan, S., Sanjay, M. R., & Siengchin, S. (2020). Suitability evaluation of sidamysorensis plant fiber as reinforcement in polymer composite. *Journal of Natural Fibers,* 1–11.

10. Hassan, T., Jamshaid, H., Mishra, R., Khan, M. Q., Petru, M., Novak, J., ... & Hromasova, M. (2020). Acoustic, mechanical and thermal properties of green composites reinforced with natural fibers waste. *Polymers, 12*(3), 654.

11. Kumar, S., Zindani, D., & Bhowmik, S. (2020). Investigation of mechanical and viscoelastic properties of flax-and ramie-reinforced green composites for orthopedic implants. *Journal of Materials Engineering and Performance, 29,* 3161–3171.

12. Prasath, K. A., Arumugaprabu, V., Amuthakkannan, P., Manikandan, V., & Johnson, R. D. J. (2020). Low velocity impact, compression after impact and morphological studies on flax fiber reinforced with basalt powder filled composites. *Materials Research Express, 7*(1), 015317.

13. Charron, J. P., Desmettre, C., & Androuët, C. (2020). Flexural and shear behaviors of steel and synthetic fiber reinforced concretes under quasi-static and pseudo-dynamic loadings. *Construction and Building Materials, 238,* 117659.

14. Liu, X., Sun, Q., Yuan, Y., & Taerwe, L. (2020). Comparison of the structural behavior of reinforced concrete tunnel segments with steel fiber and synthetic fiber addition. *Tunnelling and Underground Space Technology, 103,* 103506.

15. Binoj, J. S., Raj, R. E., Hassan, S. A., Mariatti, M., Siengchin, S., & Sanjay, M. R. (2020). Characterization of discarded fruit waste as substitute for harmful synthetic fiber-reinforced polymer composites. *Journal of Materials Science, 55*(20), 8513–8525.

16. Fallah-Valukolaee, S., & Nematzadeh, M. (2020). Experimental study for determining applicable models of compressive stress–strain behavior of hybrid synthetic fiber-reinforced high-strength concrete. *European Journal of Environmental and Civil Engineering, 24*(1), 34–59.

17. Deb, S., Mitra, N., Maitra, S., & Basu Majumdar, S. (2020). Comparison of mechanical performance and life cycle cost of natural and synthetic fiber-reinforced cementitious composites. *Journal of Materials in Civil Engineering, 32*(6), 04020150.

18. Lokesh, P., Kumari, T. S., Gopi, R., & Loganathan, G. B. (2020). A study on mechanical properties of bamboo fiber reinforced polymer composite. *Materials Today: Proceedings, 22,* 897–903.

19. Yoo, D. Y., Kim, S., Park, G. J., & Park, J. J. (2020). Residual performance of HPFRCC exposed to fire–Effects of matrix strength, synthetic fiber, and fire duration. *Construction and Building Materials, 241,* 118038.

20. Billady, R. K., & Mudradi, S. (2020, May). Influence of filler incorporation on the mechanical and wear behaviour of synthetic fiber reinforced polymer matrix composites-A review. In AIP Conference Proceedings (Vol. 2236, No. 1, p. 040001). AIP Publishing LLC.

21. Kazmi, S. M. S., Munir, M. J., Wu, Y. F., Patnaikuni, I., Zhou, Y., & Xing, F. (2019). Axial stress-strain behavior of macro-synthetic fiber reinforced recycled aggregate concrete. *Cement and Concrete Composites, 97,* 341–356.

22. Rahman, R., & Putra, S. Z. F. S. (2019). Tensile properties of natural and synthetic fiber-reinforced polymer composites. In Mohammad Jawaid, Mohamed Thariq Bin Haji Hameed Sultan, & Naheed Saba (eds.), Mechanical and Physical Testing of Biocomposites, Fibre-Reinforced Composites and Hybrid Composites (pp. 81–102). Malaysia: Woodhead Publishing. Elsiever, Netherlands.

23. Khan, M. Z. R., Srivastava, S. K., & Gupta, M. K. (2020). A state-of-the-art review on particulate wood polymer composites: Processing, properties and applications. *Polymer Testing*, *03*, 106721.

24. Oladele, I. O., Adediran, A. A., Akinwekomi, A. D., Adegun, M. H., Olumakinde, O. O., & Daramola, O. O. (2020). Development of ecofriendly snail shell particulate-reinforced recycled waste plastic composites for automobile application. *The Scientific World Journal*, 208–248.

25. Sharma, A., Kumar, S. A., & Kushvaha, V. (2020). Effect of aspect ratio on dynamic fracture toughness of particulate polymer composite using artificial neural network. *Engineering Fracture Mechanics*, *228*, 106907.

26. Bhattacharjee, A., Ganguly, K., & Roy, H. (2020). An operator based novel micromechanical model of viscoelastic hybrid woven fibre-particulate reinforced polymer composites. *European Journal of Mechanics-A/Solids*, *03*, 104044.

27. Srinivasan, T., Suresh, G., Santhoshpriya, K., Chidambaram, C. T., Vijayakumar, K. R., & Munaf, A. A. (2020). Experimental analysis on mechanical properties of banana fibre/epoxy (particulate) reinforced composite. *Materials Today: Proceedings*, Vol. 45, Part 2, 2021, pp. 1285–1289.

28. Gopan, S. N., & Ravichandran, M. (2020). Effect of chemical treatment on mechanical properties of prawn antenna reinforced waste plastic particulates filled polymer composites. *Materials Today: Proceedings*, Vol. 33, Part 7, 2020, pp. 3668–3675.

29. Surana, I., Bedi, H. S., Bhinder, J., Ghai, V., Chauhan, A., & Agnihotri, P. K. (2020). Compression and fracture behavior of leather particulate reinforced polymer composites. *Materials Research Express*, *7*(5), 054006.

30. Jeyapragash, R., Srinivasan, V., & Sathiyamurthy, S. (2020). Mechanical properties of natural fiber/particulate reinforced epoxy composites–A review of the literature. *Materials Today: Proceedings*, *22*, 1223–1227.

31. Pareta, A. S., Gupta, R., & Panda, S. K. (2020). Experimental investigation on fly ash particulate reinforcement for property enhancement of PU foam core FRP sandwich composites. *Composites Science and Technology*, 108207.

32. Prasath, K. A., Amuthakkannan, P., Arumugaprabu, V., & Manikandan, V. (2019). Low velocity impact and compression after impact damage responses on flax/basalt fiber hybrid composites. *Materials Research Express*, *6*(11), 115308.

33. Chaudhary, V., & Ahmad, F. (2020). A review on plant fiber reinforced thermoset polymers for structural and frictional composites. *Polymer Testing*, Vol. 91, November 2020, 106792.

34. Cui, X., Honda, T., Asoh, T. A., & Uyama, H. (2020). Cellulose modified by citric acid reinforced polypropylene resin as fillers. *Carbohydrate Polymers*, *230*, 115662.

35. Hemath, M., MavinkereRangappa, S., Kushvaha, V., Dhakal, H. N., & Siengchin, S. (2020). A comprehensive review on mechanical, electromagnetic radiation shielding, and thermal conductivity of fibers/inorganic fillers reinforced hybrid polymer composites. *Polymer Composites*, *03*, 1–26.

36. Paul, R., & Bhowmik, S. (2020). Tribological behavior of micro coir filler reinforced polymer composite under dry, wet, and heated contact condition. *Journal of Natural Fibers*, *06*, 1–16.

37. Omrani, E., Menezes, P. L., & Rohatgi, P. K. (2020). State of the art on tribological behavior of polymer matrix composites reinforced with natural fibers in the green materials world. *Engineering Science and Technology, an International Journal*, *19*(2), 717–736.

38. Matykiewicz, D. (2020). Biochar as an effective filler of carbon fiber reinforced bioepoxy composites. *Processes*, *8*(6), 724.

39. Jayamani, E., Nair, G. A., & Soon, K. (2020). Investigation of the dielectric properties of natural fibre and conductive filler reinforced polymer composites. *Materials Today: Proceedings*, *22*, 162–171.

40. Madhu, S., & Balasubramanian, M. (2020). Influence of seaweed filler on dry sliding wear of carbon fiber reinforced epoxy composites. *Journal of Natural Fibers, 02*, 1–11.

41. Ye, F., Zhang, X., Liao, K., Lu, Q., Zou, X., Ran, R., ... & Shao, Z. (2020). A smart lithiophilic polymer filler in gel polymer electrolyte enables stable and dendrite-free Li metal anode. *Journal of Materials Chemistry A, 8*(19), 9733–9742.

42. Drozdov, A. D., & de Claville Christiansen, J. (2020). Modeling electrical conductivity of polymer nanocomposites with aggregated filler. *Polymer Engineering & Science, 10*, 1–10.

43. Oliver-Borrachero, B., Sanchez-Caballero, S., Fenollar, O., & Sellés, M. A. (2019). Natural-fiber-reinforced polymer composites for automotive parts manufacturing. In Key Engineering Materials (Vol. 793, pp. 9–16). Switzerland: Trans Tech Publications Ltd. Switzerland.

44. Eivazzadeh-Keihan, R., Chenab, K. K., Taheri-Ledari, R., Mosafer, J., Hashemi, S. M., Mokhtarzadeh, A., ... & Hamblin, M. R. (2020). Recent advances in the application of mesoporous silica-based nanomaterials for bone tissue engineering. *Materials Science and Engineering: C, 107*, 110267.

45. Rababah, M. M., & AL-Oqla, F. M. (2020). Biopolymer composites and sustainability. In Faris Al-Oqla S.M. Sapuan (ed.), Advanced Processing, Properties, and Applications of Starch and Other Bio-Based Polymers (pp. 1–10). Netherlands: Elsevier, Netherlands.

46. Zhang, Q., Khan, M. U., Lin, X., Yi, W., & Lei, H. (2020). Green-composites produced from waste residue in pulp and paper industry: A sustainable way to manage industrial wastes. *Journal of Cleaner Production, 262*(12), 121251.

47. George, A., Sanjay, M. R., Sriusk, R., Parameswaranpillai, J., & Siengchin, S. (2020). A comprehensive review on chemical properties and applications of biopolymers and their composites. *International Journal of Biological Macromolecules*, vol. 154, 1 july 2020, 329–338.

48. Sengupta, S. (2020). Development of Jute Fabric for Jute-Polyester Biocomposite considering Structure–Property Relationship: Jute Fabric Structure for Biocomposite. *Journal of Natural Fibers, 08*, 1–15.

49. Yadav, S. K. J., Vedrtnam, A., & Gunwant, D. (2020). Experimental and numerical study on mechanical behavior and resistance to natural weathering of sugarcane leave reinforced polymer composite. *Construction and Building Materials, 262*, 120–135.

50. Kerni, L., Singh, S., Patnaik, A., & Kumar, N. (2020). A review on natural fiber reinforced composites. *Materials Today: Proceedings*, vol. 28, part 3, 2020, 1616–1621, 10–18.

51. Nayak, S. Y., Sultan, M. T. H., Shenoy, S. B., Kini, C. R., Samant, R., Shah, A. U. M., & Amuthakkannan, P. (2020). Potential of natural fibers in composites for ballistic applications – A review. *Journal of Natural Fibers, 02*, 1–11.

52. Ramasubbu, R., & Madasamy, S. (2020). Fabrication of automobile component using hybrid natural fiber reinforced polymer composite. *Journal of Natural Fibers, 04*, 1–11.

3 Design of Polymer-Based Composites

J.T. Winowin Jappes
Kalasalingam Academy of Research and Education

S. Vignesh
Kalasalingam Academy of Research and Education

K. Sankaranarayanan
Kalasalingam Academy of Research and Education

M. Thirukumaran
PSN College of Engineering and Technology

N.C. Brintha
Kalasalingam Academy of Research and Education

CONTENTS

3.1 PARAMETERS

3.1.1 Matrix

A matrix is used to bind the fibers of the reinforcement together so that stresses are distributed throughout the materials. Matrix of a polymer-based composite is divided into three subcategories, namely, thermosets, thermoplastics, and rubber. Usually, polymer matrix composite tends to be less dense than a metal or ceramic matrix. They are known to resist atmospheric corrosion, and also exhibit superior resistance to electricity conduction. Binding epoxy polyester and phenolic resins are the most commonly used binders in polymer-based composites. At elevated temperatures, the binding properties may decrease as the polymers have less resistance to heat. For activation, the surface is activated by oxidizing the polymer surface to form polar functional groups which increases the surface energy, such that they have a better chance of interacting/binding with other materials like adhesives or even paint [1–10]. The tensile strength (minimum and maximum) suggested by some researchers is shown in Figure 3.1 [12–17].

3.1.2 Reinforcement

Taking fiber into consideration, the length of the fiber, the orientation of the fiber, and the thickness of each fiber strand are determined based on the application of the polymer composite. Hence, polymers have preferred shapes and orientations;

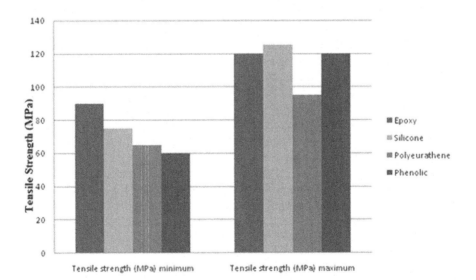

FIGURE 3.1   Tensile strength of various matrix materials.

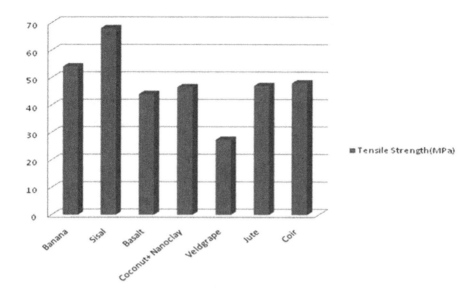

FIGURE 3.2 Tensile strength of various reinforcement materials.

they can also add more heat resistance or heat conduction of the material based on the application. A reinforcement which is added to strengthen the matrix must be stronger with less or no ductility [2,5]. On the other hand, metal powders when added as reinforcement make the distribution and dispersion crucial. Dispersion is a measure of whether the particles are agglomerated and whether the particles have filled the space homogenously. Hence, with good dispersion and a right shape and size containing no agglomerate and good distribution, each particle has equal faraway spaces with homogeneity [5,11]. Figure 3.2 shows some of the standard reinforcement materials suggested by researches as well as their tensile strength [12–17].

3.1.3 GEOMETRICAL PARAMETERS

Geometrical parameters such as thickness and curvature are some of the vital aspects that affect the impact behavior of a composite geometric structure because they are capable of altering energy absorption mode and the damage area. The toughness of the composite is influenced by thickness because the damage restriction of damage area and the threshold rises with an increase in thickness. Taking environmental conditions into consideration, we can speculate that the damage can be both permanent and reversible in polymer-based composites as well as ordinary composites. These may even lead to premature damage in the composite material. When the materials are exposed to elevated temperatures, the composite materials get residual stresses through the thickness which leads to unequal thermal contraction [18,19].

3.2 CONVENTIONAL MANUFACTURING PROCESSES

3.2.1 PREPREGS

Prepregs are obtained by combining fibers with uncured resin, which are pre-infused with thermoset resin or a thermoplastic resin matrix that requires only temperature to get activated and energized. These prepregs are readily available and the different layers are infused and cut into several pieces, and thereby can be placed and stuffed in the open mold cavity. The hand layup process is one of the prominent and promising methods of composite manufacturing in which fibers are initially spread in a mold over which a minute layer of adhesive matrix is coated for easy extraction. A brush on reinforced material is used where the resin material is poured or applied. Rollers are used to force the resin into the fabrics, which is also done to ensure an enhanced interaction between every layer of the reinforcement and matrix material [3,6,18].

3.2.2 INJECTION MOLDING

Injection molding can fabricate composite parts with high precision and exceptional low cycle times. In the injection molding process, fiber composites are fed through a hopper in the form of pellets which are then conveyed via a screw conveyor with a barrel that is subjected to heat to obtain the necessary plasticity. The material is injected through the screw via a nozzle into the cavity of the mold immediately after the needed amount of material is melted. The process is followed by cooling of the material so that the required shape and necessary properties can be obtained as a result of solidification. Injection molding is a widely proven methodology which is very efficient for making thermoplastic composites applicable in electronic products and medical applications. By improving the fiber–matrix affinity and reduction in agglomeration of fibers in the matrix material, high quality products can be obtained [6,18].

3.2.3 VACUUM BAG MOLDING

Vacuum bag molding is a process in which a flexible film made of nylon polyethylene or polyvinyl alcohol is isolated from the outside atmosphere by sealing the part. This technique is usually performed following the hand layup. The hand layup technique is used to make the laminate, which is then placed between the mold and vacuum bag to properly mix the fibers into the matrix material. Then, the vacuum pump is used to draw out the air present between the fibers and matrix at a pressure that compresses the composite. This process eliminates the probability of porosity and agglomeration of dual reinforcements. It also increases the flexural and inter-laminar shear properties [7,18,21].

3.2.4 VACUUM INFUSION PROCESS

Vacuum infusion process uses preheated molds which are mounted on either a hydraulic press or a mechanical press. A reinforcement package prepared from the prepreg is placed between the two halves of the mold, which are then pressed against each other to get the shape of the mold. The process offers a short cycle time with high productivity and automation as well as dimensional stability. Therefore, the synthesized product finds various applications in automobile industries. For example, the jute fiber-reinforced epoxy polymer composite is fabricated by hand layup, followed by the compression molding technique at curing temperatures ranging from 75°C to 130°C. Various enhancements in the mechanical properties are seen with an increase in textural impact and structural change [7,18].

3.2.5 VACUUM-ASSISTED RESIN TRANSFER MOLDING

Vacuum-assisted resin transfer molding is the process in which the preform fibers are placed on the mold and the perforated tube is positioned between the resin container and the vacuum bag. The resin is sucked through the perforated tubes because of the vacuum force, which in turn consolidates the laminate structure. This process does not leave excess air in the composite; therefore, it is used in making large objects like wind turbine blades and boat hauls. For improving the strength of the textile composites, surface-treated natural fibers are used [7,20,21].

3.2.6 SPRAY-UP TECHNIQUE

Spray-up technique uses a hand gun which is used to spray the pressurized chopped fibers, the powdered fibers, and the resin on to the mold. Then, a roller is introduced to fuse the matrix materials and the fibers together. The whole process is performed at room temperature or at elevated temperatures based on the requirement. This process is similar to hand layup process [6,18].

3.2.7 PULTRUSION

Pultrusion is a process in which the strands of fibers are pulled continuously through a resin bath and then consolidated to a heated die. Pultrusion being a continuous process is important in the fabrication of composites with a cross-section constant with longer length fiber, thereby enabling production at low cost [8,18].

3.3 THERMOPLASTIC MATRIX COMPOSITE MANUFACTURING METHODS

The processes discussed above can also be used to manufacture thermoplastic matrix composites. Processes such as tape winding, pultrusion, thermoforming, injection molding, and compression molding can be used for manufacturing thermoplastic matrix composites. Thermoforming is a method of elevating a thermoplastic sheet to a higher temperature until its softening point and then stretching it repeatedly to get a single piece of mold. Then, the material is allowed to cool so that the solidification leads to the desired shape. Thermoplastic is then clamped to the holding device and placed inside an oven. Inside the oven the material is either heated in convection or in radiation until it softens. The material then conforms to heat and converts to the desired shape. For finishing, the excess material is trimmed away, and surface finishing methods are implemented. The excess materials can be grounded and mixed and reformed into new plastic sheets [2,5,9]. Pultrusion is a continuous autonomous process, where fibers are laminated, which, in turn, produces high fiber volume profiles with a constant cross-section. Glass, carbon, and aramid materials are woven or stitched, and the resin matrix is a polyester epoxy or vinyl ester. The material is pulled to the feed area where the required shape can be accurately formed. The resin cures within the die, and then the material is allowed to cool before the pulling process. Handover reciprocation is used to pull the material. A smooth continuous pull at a constant speed is maintained to get the best result [5,8,18].

Compression molding is one of the most commonly used technique to manufacture thermoplastic composite materials, with a precise result and a rapid manufacturing rate. It is a process in which a material is molded into a confined shape by applying heat and pressure. It usually involves two stages: preheating followed by pressurizing the material. The setup consists of a fixed lower fixed mold, an upper movable mold, an ejector pin, and the charge. Compression molding is an easy process. The material is preheated to a required temperature and then placed inside the mold. The material is allowed to cure or consolidate for a certain amount of time. The material is compressed with a hydraulic press evenly to enable proper contact on all parts of the material. Due to the evenly applied pressure, the material's resin is squeezed and fills the entire mold cavity. Further, the heat present in the mold cures the material giving it the desired shape. After curing, the material is allowed to cool and carefully removed from the mold.

The whole process is done at a low cost. We can achieve uniform density, and due to uniform flow, we get uniform shrinkage in this process. There are two common types of compression molding: bulk molding and sheet molding compound. It can be used to manufacture materials like thermoplastics, thermosetting plastics, and fiber-reinforced composites.

3.4 ADVANTAGES AND DISADVANTAGES

3.4.1 ADVANTAGES

Matrix: The resin matrix forms a hard surface which protects the reinforcement material from any initial damage. Thermosets have a high dimensional stability, high temperature resistance, and good resistance to solvents. Thermoplastics have good resistance to cracking, and they also have good resistance to damage from impact [1,5].

Fibers: The fiber reinforcements are added to strengthen the materials and add stiffness to the combined material. The fibers present in the composites can be arranged in a multi-directional pattern that spreads stresses throughout the material during manufacturing [1,3,4].

Resilience: Polymer composites have good resistance to chemical corrosion, scratching, rust, and seawater. Therefore, they are used in applications such as aircraft hulls, military vehicles, trains, and boats. They also have a good wear durability [4,5].

Strength: Polymer composites have a high tensile strength-to-weight ratio which is one of the important reasons to use polymer-based composites. Composites with polyaramid fibers are five times stronger compared to steel on a pound-to-pound basis. Composite have a smooth surface once finished in aircrafts, which is useful in reducing aerodynamic drag. It is possible to achieve combinations of properties not attainable with metals or ceramics [2,5].

3.4.2 Disadvantages

The primary disadvantage of polymer composites is the process of manufacturing them into useful products. Polymer-based composites are manufactured in laboratory processes known as layup, which slows down the production rates. This also makes the products less cost-effective for high production volumes [1,11]. They also have a low compressive strength because they break easily under sudden, sharp forces. Properties of many important composites are anisotropic, and some advanced polymer-based composites are very expensive to manufacture. These advanced formulas require more expensive training for labor and more sophisticated environmental and health considerations [1,5]. Further, to shape a composite, the manufacturing cost is high and the process is not very rapid which makes this choice chaotic.

3.5   APPLICATIONS

Polymer matrix composites are extensively used in various fields such as automotive, aerospace, and marine applications. A few applications are listed below.

3.5.1 Automotive Vehicles

Tiers, belts, and hoses are made from polymer-based composites to give superior properties than the regular rubber material. On the other hand, carbon-reinforced polymer matrix composites are used to build modern expensive cars. Car components such as casing of engine parts, bumpers, fuel tanks, and body panels are made of polymer-based composites. The paintwork and windscreen glasses also use polymer-based coatings owing to the surface integrity of polymer coatings [10].

3.5.2 Aerospace Vehicles

Polymer matrix composites are used in aircraft tires and interiors, which enhances their performance and reduces the weight of the aircraft. They also provide aerospace vehicles high toughness, high strength, high stiffness, and low density. Exceptional strength-to-density and stiffness-to-density ratios result in superior physical properties of the aerospace vehicles.

3.5.3 Marine Vehicles

Fiber glass is used to make fiber glass boats which are popular nowadays. A glass fiber reinforced with a matrix polymer is arranged in different orientations like woven fabric, chopped stranded mat, or even randomly. Nowadays, carbon fibers have started to replace glass fibers to give superior properties to the products [10,18].

3.5.4 Medical Devices

Magnetic resonance imaging scanners, computed tomography scanners, mammography plates and tables, X-ray couches, surgical tools, prosthetics, and wheel chairs are some of the well-known polymer-based composites. Polymer matrix nanocomposites made of carbon nanotubes or titanium oxide nanotubes are known to reduce the inherent wound binding time of fractured bones by acting as a scaffold, thus paving the way for the development of replacement bone.

3.5.5 Personal Guards

Polymer matrix composites are used in helmets, gloves, boots, glasses, and cloths as they have superior properties to normal materials and can withstand more impact and abrasion. They protect well against some of the conditions like radiation, explosion, electricity, and heat. Composites have more benefits than regular materials which results in their extensive use in personal protective equipment. Football boots, sports shoes, gloves, helmets, and rackets are some of the application where polymer matrix composites are widely used. They are also used in industrial equipment and in packaging. Polymer matrix composites are also used to repair and replace a structural component such as bricks. Polymer-based composites are used in blades of windmills. These windmill blades achieve extreme strength-to-weight ratio by utilizing polymer matrix composites to the fullest, which is necessary for boosting the output. Polymer matrix nanocomposites are used in thin film charge storage devices called capacitors for the internal motherboard and storage devices in the form of computer chips.

REFERENCES

1. Abdi B, Azwan S, Abdullah MR, Ayob A, Yahya Y Xin L. Flatwise compression and flexural behavior of foam core and polymer pinreinforced foam core composite sandwich panels. *International Journal of Mechanical Sciences*, 88 (2014), 138–144.
2. Adekunle K, Cho SW, Patzelt C, Blomfeldt T, Skrifvars M. Impact and flexural properties of flax fabrics and Lyocell fiber-reinforced biobased thermoset. *Journal of Reinforced Plastics and Composites* (2011). doi:1177/0731684411405874.
3. Ahmed KS, Vijayarangan S. Tensile, flexural and interlaminar shear properties of woven jute and jute-glass fabric reinforced polyester composites. *Journal of Materials Processing Technology*, 207(1–3) (2008), 330–335.
4. Aji IS, Zainudin ES, Khalina A, Sapuan SM, Khairul MD. Studying the effect of fiber size and fiber loading on the mechanical properties of hybridized kenaf/PALF-reinforced HDPE composite. *Journal of Reinforced Plastics and Composites*, 30(6) (2011), 546–553.
5. Alavudeen A, Rajini N, Karthikeyan S, Thiruchitrambalam M, Venkateshwaren N. Mechanical properties of banana/kenaf fiberreinforced hybrid polyester composites: Effect of woven fabric and random orientation. *Materials & Design*, 66 (2015), 246–257.
6. Abdelwahab MA, Flynn A, Chiou BS, Imam S, Orts W, Chiellini E. Thermal, mechanical and morphological characterization of plasticized PLA–PHB blends. *Polymer Degradation and Stability*, 97 (2012), 1822–1828.
7. Al-Oqla FM, Omar, AA. A decision-making model for selecting the GSM mobile phone antenna in the design phase to increase overall performance. *Progress in Electromagnetics Research C*, 25 (2012), 249–269.

8. AL-Oqla FM, Sapuan, SM. Natural fiber reinforced polymer composites in industrial applications: feasibility of date palm fibers for sustainable automotive industry. *The Journal of Cleaner Production*, 66 (2014), 347–354.
9. AL-Oqla FM, Sapuan SM, Ishak MR, Nuraini AA. A novel evaluation tool for enhancing the selection of natural fibers for polymeric composites based on fiber moisture content criterion. *BioResources*, 10(1) (2014), 299–312.
10. AL-Oqla FM, Sapuan SM, Ishak MR, Nuraini AA. Selecting natural fibers for bio-based materials with conflicting criteria. *American Journal of Applied Sciences*, 12(1) (2015), 64–71.
11. Alavudeen A, Thiruchitrambalam M, Venkateshwaran N, Athijayamani A. Review of natural fiber reinforced woven composite. *Review of Advanced Materials Science*, 27 (2011), 146–150.
12. Mayandi K, Rajini N, Pitchipoo P, WinowlinJappes JT, Siva I. Mechanical performance of Cissus quadrangularis/polyester composite. *Materials Today Communications*, 4 (2015), 222–232.
13. Amuthakkannan P, Manikandan V, Winowlin Jappes JT, Uthayakumar M. Effect of fibre length and fibre content on mechanical properties of short basalt fibre reinforced polymer matrix composites. *Materials Physics and Mechanics*, 16 (2013), 107–117.
14. Senthil Kumar K, Siva I, Jeyaraj P, Winowlin Jappes JT, Amico SC, Rajini N. Synergy of fiber length and content on free vibration and damping behavior of natural fiber reinforced polyester composite beams. *Materials and Design*, 56 (2014), 379–386.
15. Manikandan V, Winowlin Jappes JT, Suresh Kumar SM, Amuthakkannan P. Investigation of the effect of surface modifications on the mechanical properties of basalt fibre reinforced polymer composites. *Composites: Part B*, 43 (2012), 812–818.
16. Rajini N, Winowlin Jappes JT, Rajakarunakaran S, Jeyaraj P. Dynamic mechanical analysis and free vibration behavior in chemical modifications of coconut sheath/nano-clay reinforced hybrid polyester composite. *Journal of Composite Materials*, 47(24) (2013), 3105–3121.
17. Neser G. Polymer based composites in marine use: History and future trends. *Procedia Engineering*, 194 (2017), 19–24.
18. Alvira P, Tomas-Pejo E, Ballesteros M, Negro MJ. Pretreatment technologies for an efficient bioethanol production process based on enzymatic hydrolysis: A review. *Bioresource Technology*, 101 (2010), 4851–4861.
19. Ashok Kumar M, Ramachandra Reddy G, Harinatha Reddy G, Chakradhar KVP, Nanjundareddy BH, Subbarami Reddy N. Mechanical properties of randomly oriented short sansevieria trifascatafiber/epoxy composites. *International Journal of Fiber and Textile Research*, 1 (1) (2011), 6–10.
20. Haneefa A, Bindu P, Aravind I, Thomas S. Studies on tensile and flexural properties of short banana/glass hybrid fiber reinforced polystyrene composites. *Journal of Composite Materials*, 42(15) (2008), 1471–1489.
21. Ashok Kumar M, Ramachandra Reddy G, Vishnu Mahesh KR, Hemanth Babu T, Vasanth Kumar Reddy G, Dasaratha H, Mohana Reddy Y V. Fabrication and performance of natural fibers: Sansevieria cylindrica, waste silk, jute and drumstick vegetable fibres (Moringa Oleifera) reinforced with rubber/polyester composites. *International Journal of Fiber and Textile Research*, 1 (2011), 15–21.

4 Design of Polymer Hybrid Composites

T. Sathish

Saveetha School of Engineering, SIMATS

N. Sabarirajan

Chendhuran College of Engineering and Technology

CONTENTS

4.1 INTRODUCTION

Composite is the structural combination of two or more materials at the macroscopic level in which selected constituents are insoluble. One of the constituents in the composite is called as the reinforcing material or phase and another acts as the matrix or base material (Espinosa et al., 2009). The reinforcement materials are used to change the characteristics of the base material, and can be in the form of flakes, fibres or particles. These material matrix compositions are widely used in industries to change or enhance the mechanical characteristics, especially in terms of strength (Shirvanimoghaddam et al., 2017).

In automotive industries, as aluminium plays a major role, in matrix composition, aluminium is considered as the base material. Aluminium-based composition is also widely used in aerospace industries for manufacturing of aircraft components. For making hard parts or components like piston rings, brake rotors, etc., ceramic reinforcements are used (Yu et al., 2019). Previous studies have shown that these compositions enhance the tribological properties along with the mechanical property.

Even though the mechanical properties of these composites are enhanced, the composite becomes heavy due to the ceramic reinforcement in aluminium matrix. Hence, an alternative to aluminium metal is required to reduce the weight of the composite (Kumar et al., 2020).

As a solution, some researchers have tried to use magnesium as an alternative to aluminium. Magnesium has less density than aluminium, so it acts as a lightweight composite with enhanced strength and damping capacity (Lu et al., 2009). Hence, in a short span, magnesium has replaced aluminium in aerospace and automobile parts manufacturing. Because lightweight composites using magnesium are expensive, researchers are trying to invent a low-cost composite. One solution is to use polymer-based composite for better performance (Rana et al., 2012).

Hence, lately, engineering research has shifted to polymeric materials-based composite preparation from conventional monolithic materials (Meng et al., 2017). Natural glass fibre reinforcement is one of the best alternatives for monolithic materials. In polymer composite manufacturing, natural fibre reinforcement provides better performance at affordable prices (Roberts et al., 2014). Many recent researchers are attracted towards the usage of natural fibre instead of carbon or conventional glass fibres. Some of the most used natural fibres are hemp, flax, sisal, jute, coir, kenaf, banana, kapok and so on. These naturally available fibres have better mechanical properties in terms of modulus, stiffness, as well as flexibility (Holbery and Houston, 2006).

Some of the noticeable advantages of natural fibre reinforcement are (Lee et al., 2005):

- Low density and cost.
- Ability to achieve specific tensile properties.
- Non-abrasive in nature to the equipment.
- Mostly non-irritation to the body or skin.
- Consume less energy, and low health risk.
- Recyclable, renewable and biodegradable in nature.

Some of the conventional glass or synthetic fibre reinforcements can produce better strength, but these composites have limited usage due to their production cost. On the other hand, natural fibres can be strong in terms of cost and weight, and are better than synthetic and glass reinforcements.

Among natural fibres, jute and sisal fibres are easily available; hence, in most composites, these fibres are used for reinforcement instead of synthetic fibres. The usage of natural fibres can also help to reduce environmental hazards because these fibres are biodegradable in nature. Most of the time the choice of natural fibres is based on availability in the local region. Moreover, natural fibres have better mechanical properties in composites (Coutts, 2005).

In specific applications, the overall strength required by the composite material might not be met by the natural fibre reinforcement. In these compositions, the single fibre reinforcement has been made. Therefore, researchers tend to use more than one reinforcement to make hybrid composites. In some hybridisation processes, glass and fibre are mixed at a specific proportion to make hybrid reinforcement. The hybrid polymer composite can enhance the properties at an affordable price. The fibre-reinforced polymer (FRP) is a kind of matrix composite in which a polymer acts as the base metal and fibre material is used as the reinforcement. In this chapter, we discuss the hybrid reinforcement in polymer matrix composite preparation.

4.2 HYBRID COMPOSITES TYPES

In FRP, polymer composites are prepared with fibre reinforcement. Hybrid reinforcements are used to improve mechanical properties. In hybrid composites, more than one reinforcement is used to attain the desired properties (Killick, 2019). Based on the selection of fibre type for the composition, the hybrid composite can be classified into three types such as:

1. Synthetic-synthetic fibre composition
2. Natural-natural fibre composition
3. Synthetic-natural fibre composition

Figure 4.1 shows the classification of different types of fibres. Based on the application, one of the three hybridisation technique is preferred. To attain the optimal performance, the synthetic-natural composition is used.

The classification of fibres is shown in Figure 4.1. Fibres are divided into two main categories, namely, synthetic and natural. Synthetic fibres are subdivided into two categories, namely, organic and inorganic fibre. Natural fibres have three subcategories.

4.2.1 Synthetic-Synthetic Fibre Composition

The synthetic-synthetic fibre composition combines two synthetic fibres to make hybrid composites. Glass, carbon and boron fibres are the most used synthetic fibres in hybrid matrix composites, which are shown in Figure 4.2. The hybrid polymer composites using synthetic fibre reinforcements are widely used for manufacturing automotive and aerospace components. The properties of the commonly used synthetic fibres are listed in Table 4.1 (Asim et al. 2017).

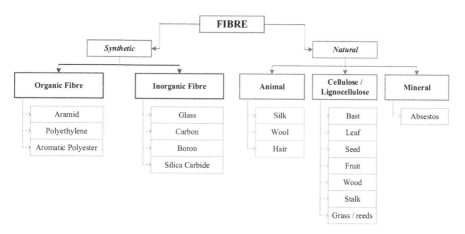

FIGURE 4.1   Classification of fibre.

Glass   Carbon   Boron

FIGURE 4.2   Synthetic fibres.

TABLE 4.1
Properties of Synthetic Fibres

Properties	Unit	Fibre		
		Glass	**Carbon**	**Boron**
Tensile strength	MPa	3,445	4,000	3,600
Compressive strength	MPa	1,080	869	6,000
Density	g/cm³	2.58	2.267	2.61
Thermal expansion	μm/m°C	5	–	4.5
Softening temperature	°C	846	3,652	2,040

Few synthetic fibres sample images are shown in Figure 4.2, and the most used synthetic fibres such as glass, carbon and boron are displayed.

The strength property of boron is comparatively higher than all other materials. Similarly, other properties such as density and softening temperature are also better in boron than glass and carbon.

4.2.2   NATURAL-NATURAL FIBRE COMPOSITION

The natural-natural fibre composition combines two natural fibres for reinforcement in polymer matrix composite. The most used natural fibres are wood, sisal, hemp, coconut, cotton, kenaf, flax, jute, abaca, banana leaf fibres, bamboo and wheat straw. These natural composites are cost-effective, lightweight and have low energy production. Some of the most used natural fibres are shown in Figure 4.3, and their properties are listed in Table 4.2.

4.2.3   SYNTHETIC-NATURAL FIBRES COMPOSITES

The synthetic-natural reinforced hybrid polymer composite combines a synthetic and natural fibre material. This hybrid combination of fibre can help obtain the optimal performance at low cost. Here, for the composition, different weight, chemical composition and mechanical properties are applied on the same fabric or composition. This type of composition is used to attain a specific property of component at low cost. For example, to attain diverse elastic properties, a mixture of carbon and aramid is combined to make the matrix composite. In this type of composition,

FIGURE 4.3 Natural fibres.

TABLE 4.2
Properties of Natural Fibres

	Properties		
	Tensile Strength	Density	Softening Temperature
Fibre	MPa	g/cm³	°C
Wood	25	1.04	227
Sisal	353	1.24	210
Hemp	250	0.86	420
Coconut	225	0.67	120
Cotton	44	1.54	246
Kenaf	780	1.2	113
Flax	1800	1.42	219
Jute	700	1.5	125
Abaca	190	1.5	173
Banana	62	1.35	168
Bamboo	810	0.9	125
Wheat straw	2585	0.3	190

the proportion of synthetic fibre should be low compared to the proportion of natural fibre. Because natural fibres are cheaper than synthetic fibres, the overall cost of composite is reduced.

4.3 HYBRID COMPOSITES PROCESSING

The natural fibre composition is used to develop a biodegradable product and to manufacture low-cost and strong components for industrial applications. In this section, the process of hybrid polymer composite using natural fibre reinforcement is presented (Cheung et al., 2009). There are two main methods used for the preparation of hybrid polymer composites:

1. Pultrusion
2. Injection moulding

4.3.1 PULTRUSION

Pultrusion is the process of polymer composite preparation which combines pull and extrusion processes. The pultrusion process is similar to that of the extrusion process, but in pultrusion, the materials are poured into mould state for developing the component or product. The process flow of pultrusion mechanism is given in Figure 4.4.

In the pultrusion process, the raw fibre is fed to the resin bath to make the resin structure on the fibre, which is then shaped using a shaper or die. Then, a pultrusion die is used for pultrusion and a puller is used to pull the model and cut-off to segregate the product.

4.3.2 INJECTION MOULDING

Injection moulding is one of the widely used techniques in manufacturing, especially plastic components. The natural fibre-reinforced composites can also be prepared using injection moulding, but the chosen reinforcement material should be powdered. A twin-screw extruder is used in injection moulding to prepare granules. The process flow diagram of injection moulding is shown in Figure 4.5.

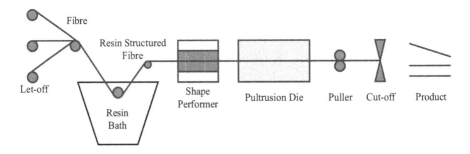

FIGURE 4.4 Pultrusion process flow.

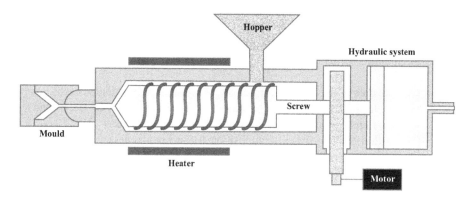

FIGURE 4.5 Injection moulding.

The process includes a hydraulic system to apply pressure to the screw to melt to the mould and create a product.

The FRP composite can also be made using some other methods such as compression moulding, vacuum infusion moulding, hot press processes and resin transfer moulding.

4.4 ADVANTAGES AND DISADVANTAGES

The hybrid FRP composite has many advantages and disadvantages based on the choice of materials and manufacturing process (Kuzman et al., 2013). Some of them are listed below.

4.4.1 ADVANTAGES

The hybrid polymer composition has many advantages, and some of the noticeable advantages of various processes and hybrid composition are listed below:

- The natural fibre composition for reinforcement provides more benefits in polymer composition, such as low cost and better performance in terms of mechanical properties.
- The hybrid fibre reinforcement can help attain improved characteristics of composites, such as ultimate tensile strength and density.
- The injection moulding process for the composite preparation has low material wastage and the leakage flow is negligible.
- Resin transfer moulding is a better technique for polymer composite preparation and helps to attain desired quality and performance.
- The resin transfer method helps to prepare very large and complex shapes of product, with better design flexibility.
- Compression moulding also has advantages in terms of low cost, short cycle time, high-volume production, dimensional accuracy, improved impact strength and uniform shrinkage.

4.4.2 DISADVANTAGES

The hybrid polymer composition has many advantages due to the usage of natural fibres and manufacturing mechanisms. Some of them are listed below:

- The hydrophilic nature of natural fibres leads to microbial degradation.
- In hybridisation, due to the hydrophilic nature, it might not create a bond between fibres, which reduces the strength of the composite.
- The natural fibre in hybrid composition has a homogeneous distribution, which can weaken the mechanical strength.
- The lignocellulosic nature of natural fibre is highly sensitive to temperature. At varying temperatures, it affects the mechanical and chemical properties.
- The injection moulding process can create some problems in Warpage, resulting in stress crack and shrinkage.
- The handling of resin transfer moulding is difficult as it requires an expert support for manufacturing.

4.5 APPLICATIONS

The hybrid polymer composites are widely used in various industrial applications. The application of hybrid polymer composites is varied based on the materials and processing method. Some of the noticeable advantages of hybrid polymer composites are listed in this section. The hybrid polymer composites using natural fibre reinforcement are used in automobile industries. These natural fibre-reinforced composites are used for manufacturing of both interior and exterior parts of automobiles. An application of natural FRP composite is shown in Figure 4.6.

Alt Text for the Figure 4.6

Figure 4.6, (Peças et al., 2018) shows the conversion of raw natural fibre (hemp) to a part of automobile.

This composite has advantages such as very light in weight ratio and good mechanical strength. Moreover, these hybrid polymer composites are also used in construction and furniture. Thus, fibre-reinforced hybrid polymer composites have several applications in today's technological world.

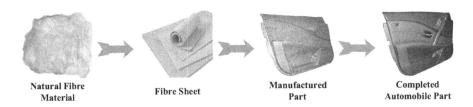

| Natural Fibre Material | Fibre Sheet | Manufactured Part | Completed Automobile Part |

FIGURE 4.6   Application of natural fibre-reinforced polymer composite.

REFERENCES

Asim,M., Jawaid, M., Saba, N., Nasir, M., and Sultan, M.T.H. 2017. Processing of hybrid polymer composites—A review. In Thakur, V. K., Thakur, M. K., and Gupta, R. K. (eds.),
Hybrid Polymer Composite Materials. India: Woodhead Publishing, 1–22.

Chand, N., and Fahim, M. 2008. *In Woodhead Publishing Series in Composites Science
and Engineering, Tribology of Natural Fiber Polymer Composites*. United Kingdom:
Woodhead Publishing, 1–58.

Cheung, H.Y., Ho, M.P., Lau, K.T., Cardona, F., and Hui, D. 2009. Natural fibre-reinforced
composites for bioengineering and environmental engineering applications. *Composites
Part B: Engineering*, 40: 655–663.

Coutts, R.S. 2005. A review of Australian research into natural fibre cement composites.
Cement and Concrete Composites, 27: 518–526.

Espinosa, H.D., Rim, J.E., Barthelat, F., and Buehler, M.J. 2009. Merger of structure and
material in nacre and bone-Perspectives on de novo biomimetic materials. *Progress in
Materials Science*, 54: 1059–1100.

Holbery, J., and Houston, D. 2006. Natural-fiber-reinforced polymer composites in automotive
applications. *Jom*, 58: 80–86.

Killick, E. 2019. Hybrid houses and dispersed communities: negotiating governmentality and living well in Peruvian Amazonia. *Geoforum*. https://doi.org/10.1016/j.geoforum.2019.08.003

Kumar, R. P., Periyasamy, P., Rangarajan, S., and Sathish, T. 2020. League championship optimization for the parameter selection for Mg/WC metal matrix composition. *Materials
Today: Proceedings*, 21: 504–510.

Kuzman, M. K., Grošelj, P., Ayrilmis, N., and Zbašnik-Senegačnik, M. 2013. Comparison of
passive house construction types using analytic hierarchy process. *Energy and Buildings*,
64: 258–263.

Lee, S.M., Cho, D., Park, W.H., Lee, S.G., Han, S.O., and Drzal, L.T. 2005. Novel silk/poly
(butylene succinate) biocomposites: the effect of short fibre content on their mechanical
and thermal properties. *Composites Science and Technology*, 65: 647–657.

Lu, H., Wang, X., Zhang, T., Cheng, Z., and Fang, Q. 2009. Design, fabrication, and properties
of high damping metal matrix composites—A review. *Materials*, 2: 958–977.

Meng, Q., Cai, K., Chen, Y., and Chen, L. 2017. Research progress on conducting polymer
based supercapacitor electrode materials. *Nano Energy*, 36: 268–285.

Peças, P., Carvalho, H., Salman, H., and Leite, M. 2018. Natural fibre composites and their
applications: A review. *Journal of Composites Science*, 2: 66.

Rana, R.S., Purohit, R., and Das, S. 2012. Reviews on the influences of alloying elements on
the microstructure and mechanical properties of aluminum alloys and aluminum alloy
composites. *International Journal of Scientific and Research Publications*, 2: 1–7.

Roberts, A.D., Li, X., and Zhang, H. 2014. Porous carbon spheres and monoliths: morphology control, pore size tuning and their applications as Li-ion battery anode materials.
Chemical Society Reviews, 43: 4341–4356.

Shirvanimoghaddam, K., Hamim, S.U., Akbari, M.K., Fakhrhoseini, S.M., Khayyam, H.,
Pakseresht, A.H., and Davim, J.P. 2017. Carbon fiber reinforced metal matrix com-
posites: Fabrication processes and properties. *Composites Part A: Applied Science and
Manufacturing*, 92: 70–96.

Yu, W.H., Sing, S.L., Chua, C.K., Kuo, C.N., and Tian, X.L. 2019. Particle-reinforced metal
matrix nanocomposites fabricated by selective laser melting: A state of the art review.
Progress in Materials Science, 104: 330–379.

5 Biocomposites Based on Polymers

P. Sivaranjana
Kalasalingam Academy of Research and Education

CONTENTS

5.1 INTRODUCTION

Biocomposites are generally biodegradable, biocompatible, and ecofriendly materials. Composite is a hybrid material formed by the combination of two or more different types of materials. The hybrid material usually possesses the properties of all the materials added to it. If the combination includes a material derived from natural resources and the hybrid material developed is biodegradable, then the resultant material is known as a biocomposite. Generally a composite consists of a matrix and a filler, and if a polymer material forms the matrix/filler of a composite, then it is said to be a polymer composite. If the polymer is derived from natural resources and is biodegradable, then it is said to be a biocomposite polymer. The biodegradability factor is a major concern in developing biocomposite materials. The emergence of environment preservation has led the researchers to work toward the development of biodegradable materials. The utility of polymer (synthetic)-based materials has become unavoidable in the current lifestyle. Therefore, these polymer (synthetic)-based utility materials are being replaced with polymer biocomposites. Synthetic fibers like carbon and glass fibers are being replaced with lignocellulosic fibers such as flax, hemp, kenaf, henequen, banana, oil palm, and jute [1]. These natural fiber-reinforced biocomposites are employed in automobile, packaging, aerospace, and construction industries, where the weight needs to be reduced and efficiency needs to be increased. Natural fibers are resistant to corrosion and fatigue and are lightweight and biodegradable; however, they lack in mechanical properties and thermal resistance [2]. The mechanical property of the biocomposite depends on the adhesion between the reinforcement fiber and the matrix. Natural fibers are polar and hydrophilic in nature, whereas the matrix polymers are polar and hydrophobic, the weak interaction between them leads to the lack of mechanical properties and limits their industrial applications [3]. The weak interfacial bonding between the natural fiber and the polymer matrix can be improved by subjecting the filler material to various treatments, such as chemical treatment (acid and alkali treatment), surface modifications, and plasma treatment or even by functionalization/nanoparticle deposition [4]. The ultimate aim for developing such materials is to replace the existing conventional (petrochemical-based) materials which are causing pollution problems during manufacturing and disposal. Municipal landfills are the major contributors for the disposal of solid waste; biocomposites can be disposed with ease in such landfills. Biocomposites are developed with a variety of combinations of materials, both natural as well as synthetic in origin. It includes polymers (natural as well as synthetic) polysaccharides, proteins, sugars, metals, metal nanoparticles, ceramics, agro-wastes, and so on. These biocomposites are developed in the form of films, membranes, coatings, particles, foams, moldings, and so on [5]. Biocomposites are ecofriendly, low cost, low density, biocompatible, and can be developed by green methods. They find wide applications in the field of packaging, automobiles, biomedical appliances, etc.

5.2 TYPES OF POLYMER-BASED BIOCOMPOSITES

5.2.1 MATRIX

Polymer-based biocomposites can be classified based on the polymer (matrix) materials as well as on the types of reinforcements used. The matrix polymer can be classified into two types based on its origin as natural and synthetic (both are biodegradable). A brief classification of matrix based on the origin is given in Figure 5.1.

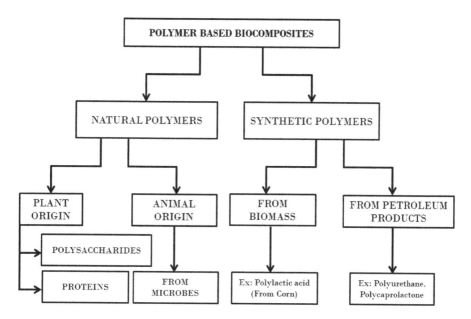

FIGURE 5.1 Classification of polymer-based composites based on the origin of the matrix material.

5.2.2 FILLERS/REINFORCEMENT

The reinforcement materials (fibers) are also classified based on their origin as natural and synthetic. Natural fibers can be further classified based on their source of origin as from plants, from animals, and from minerals [6]. Natural fibers from plants include flax, banana, napier grass, coconut sheath, aloe vera, jute, hemp, pineapple, sisal, etc. Their chemical composition includes cellulose, hemicellulose, and lignin. Natural fibers from animals include wool, silk, etc. They are composed of proteins. The hydrogen bonding in protein structures, hydrophobic nature, and their crystallinity gives extraordinary stability to protein fibers. Lignocellulosic fibers are most widely used and their properties are more promising. The adhesions of these lignocellulosic fibers with polymer matrix are improved by chemical treatments and surface modifications.

Natural fiber fillers can be classified into three categories based on the requirement of processing methodologies as straw fibers, wood fibers, and non-wood fibers. Among these, non-wood fibers find applications in most industrial sectors owing to their mechanical and physical properties. These natural fibers can be obtained from various parts of the plants such as bast, leaf, seed, fruit, and stalk [7].

Jute is one of the fibers from plant origin, which is biodegradable, affordable, can be blended with other fibers, and can be spun into lengthy, strong thread. It possesses high tensile strength, low thermal conductivity, and acoustic insulating property. Biocomposites with jute fibers are manufactured using press molding technique.

Hemp is a soft, durable fiber obtained from plants. Its Young's modulus is comparable with glass fiber. It is usually employed as an additive with polypropylene by compression molding, injection molding, and hand layup or even hybrid technologies.

Kenaf is plant fiber notable for its low density and non-abrasiveness during process-ing, appreciable mechanical property, and biodegradability. It is employed as a substitute for fiberglass and synthetic fibers. It is processed by injection molding and extrusion.

The sisal fiber is notable for its tensile strength and stiffness. It is used as a replace-ment for glass fiber. The coir from coconut is thick, strong, and abrasion resistance. The coir from coconut tree has promising properties like abrasion resistance, stronger, affordable, and economic.

5.2.3 RESINS

Resins are binding agents that hold the filler/reinforcement within the matrix. Resins are of two types based on their origin: (i) Natural resin – natural secretions from plants; (ii) Synthetic resin – both thermo and thermosetting polymers are used. To avoid environmental deterioration issues, nowadays more emphasis is given toward the utilization of natural resins. Natural resins are water resistant and soluble in organic solvents such alcohol and ether etc. They are extensively used in food, construction, and transportation industries.

5.3 BIOCOMPOSITES PROCESSING

The biocomposite processing consists of three components, namely, matrix (polymer), filler (fiber), and resin (binder). The processing methodology varies with thermoplas-tics and thermosetting plastics. The thermosetting resins can be processed either through open mold or closed mold methods. Open mold processes include hand lami-nating, spray-up, filament winding, and pultrusion. Closed mold processes include vacuum bag, autoclave, compression molding, resin transfer molding (RTM), and vacuum-assisted resin injection [8].

5.3.1 OPEN MOLD PROCESSES

5.3.1.1 Hand Laminating Method

Hand laminating is a simple and open molding type method for the fabrication of biocomposites (Figure 5.2). It involves four steps: (i) mold preparation, (ii) gel coating, (iii) layup, and (iv) curing. The chosen mold is first coated with a gel to avoid sticking

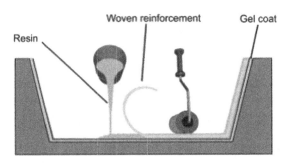

FIGURE 5.2   Hand laminating method. (Sciencedirect.com)

of polymer to the mold surface. The natural fibers (dry) are then obtained in any of the following forms: woven, stitched, or bond fabric. These fibers are then placed manually in the desired mold; the resin and matrix are added and evenly spread using a brush/hand roller in the wet condition to achieve uniform distribution and thickness. Then the mold with laminates is subjected to the curing process. Curing is a process of hardening the fiber-reinforced composite without external heat [9].

5.3.1.2 Spray-Up

In this method, both resin and fiber are added together into the mold. The reinforcements are cut and added to the mold along with resins. The products manufactured in this process are caravan bodies, truck fairings, bath tubes, small boats, etc.

5.3.1.3 Filament Winding

This is an open molding process employed in manufacturing cylindrical products like storage tanks, pipe lines, pressure vessels, and rocket motor cases. The process results in the manufacture of products with high tensile strength, high strength-to-weight ratio, and uniform fiber orientation. Filament winding is mainly used to manufacture highly engineered structures with high tolerance efficiency. The entire process is automated and requires low manpower than other open molding processes (Figure 5.3).

The filament winding process involves two components, namely, a stationary rotating steel mandrel and a carriage arm traveling horizontally up and down along the length of the mandrel. As the mandrel turns, the wrap moves around it and forms

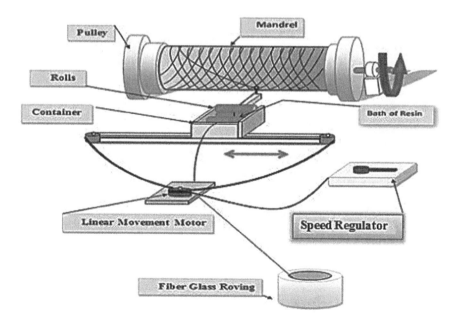

FIGURE 5.3 Filament winding machine. (https://www.researchgate.net/profile/Ehsan_AlAmeen/publication/319528444/figure/fig1/AS:535931749715968@1504787494498/Figure-1-Basic-filament-winding-process.png)

a composite layer over its surface. The travel rate of the carriage and speed of rotation of the mandrel determine the orientation of the composite matrix. The fiber impregnated with resin is fed into the rotating mandrel, filament is laid over the mold, and finally, subjected to curing in mandrel [10].

5.3.1.4 Pultrusion
In this method, continuous fiber strands are fed through a resin bath, and then passed over a heated die that cures the resin. This method is used to manufacture long beams and other continuous sections (Table 5.1).

5.3.2 CLOSED MOLD PROCESS

5.3.2.1 Vacuum Bag
In this process, reinforcement/filler and resin are mixed and applied to a mold by hand lamination techniques. A thin film or rubber bag was used to cover the laminate. The system is evacuated to apply atmospheric pressure over the surface of the composite. The applied pressure facilitates the impregnation of fiber by the resin inside the laminate. Additional hand rolling is also given to ensure the penetration of fibers. A similar process is carried out by applying pressure additional to that of the atmospheric pressure called the pressure bag method.

TABLE 5.1
Comparison of Open Mold Processes

Process	Hand Laminating	Spray Up	Filament Winding
Volume fraction	0.13–0.5	0.13–0.21	0.55–0.7
Size range	0.25–2,000 m^2	2–100 m^2	0.1–1,000 m^2
Pressure	Ambient	Ambient	Ambient
Temperature	Ambient	Ambient	Ambient
Resins	Polyester, epoxy, vinyl ester, phenolic	Polyester, vinyl ester	Polyesters, epoxy, vinyl ester
Advantages	• Low capital outlay • Secondary bonding • No size limit • Flexibility	• Low material cost • High production rate • Low tooling cost • Large parts	• Excellent mechanical properties • High production rate • Good control of fiber orientation • Good thickness control • Good fiber content control • Good internal finish
Disadvantages	• Operator dependent • Labor-intensive • Low production rate • Poor weight and thickness control • Only one molded face	• Very operator dependent • Very poor thickness control • Only one molded face • Random reinforcement only	• Limited range of shapes • Limited number of practical winding patterns

5.3.2.2 Autoclave

It is similar to that of the vacuum bag assembly. The laminate is placed inside a vacuum, and external pressure and heat are applied so that better impregnation of reinforcement and curing occurs.

5.3.2.3 Compression Molding

Compression molding is preferred in case of thermosetting polymers and is applicable for large-scale production. It is of two types: (i) hot compression and (ii) cold compression methods. The thermosetting materials are generally granular, free flowing, and viscous. The raw materials are placed on the hot mold (in case of hot compression) and the mold is closed by hydraulic pressure. A typical hot compression molding is done at a temperature of 450 K and a pressure of about 100 psi [11]. In case of cold press molding, the above process is carried out at room temperature.

5.3.2.5 Resin Transfer Molding (RTM)

RTM is similar to compression molding but the resin is added via injection ports into the mold after closure.

5.3.2.6 Vacuum-Assisted Resin Injection

In this method, the reinforcement is kept inside the mold similar to the RTM process without resin. The resin is added through injection ports under vacuum condition.

5.4 ADVANTAGES AND DISADVANTAGES

5.4.1 Advantages of Biocomposites

- **Design Flexibility** – Biocomposites offer a wide degree of flexibility in molding into various forms. They can be molded into several complicated components, with different densities and chemical formulations to achieve the desired performance properties.
- **Economic** – Biocomposites have lower material costs than traditional materials such as wood, engineered thermoplastics, and metals because they have low petroleum-based content.
- **Low processing cost** – As biocomposites are precisely molded, there is minimum wastage, and therefore, the processing charges and material costs are minimized.
- **Improved productivity** – Flexibility in molding reduces the assembling cost. There is no need of additional machining in case of biocomposites, therefore it reduces processing time as well.
- **Weight minimization** – Use of natural fiber is a revolution in manufacturing lightweight products. Products with reduced weight reduce resource usage and lower fuel consumption [12].
- They are obtained from renewable resources.
- Easy availability.
- Low density (specific weight) ensures high specific hardness and stiffness in comparison with petroleum-based materials.

TABLE 5.2
Comparison of Closed Mold Processes

Process	Vacuum Bag	Autoclave	Cold Press Molding	Hot Press Molding	Resin Transfer Molding	Vacuum-assisted Resin Injection
Volume fraction	0.15–0.6	0.35–0.7	0.15–0.25	0.12–0.4	0.1–0.15	0.15–0.35
Size range	0.5–20 m²	0.25–5.0 m²	0.25–5.0 m²	0.1–2.5 m²	0.25–5 m²	1.0–30 m²
Pressure	1 bar	Up to 10 bar	2–5 bar	50–150 bar	Max. 2 bar	Max. 2 bar
Temperature	Ambient	140°C	20°C–50°C	130°C–150°C	20°C–50°C	15°C–3°C
Resins	Most types	Epoxy	Polyester	Polyester, epoxy, vinyl ester	Polyester, polyurethane, low viscosity epoxy	Polyester, polyurethane
Advantages	• Low capital outlay. • Low cost of tooling. • Large components. • Well suited to making • Sandwich panels.	• Very high quality • High fiber content • Low void content • Controlled cure	• Good surface on both sides • Good production rate • Accurate dimensions	• Very high production rate • Fine detail, close tolerance • Low cost • Long tool life	• Good surface on both sides • Accurate dimensions • Wide range of part geometry • Reasonable production rate • Sandwich construction possible	• High fiber content • Large size • Low weight, low cost tooling • Full range of reinforcement
Disadvantages	• Labor-intensive • Low production rate. • Only one accurate surface	• Labor-intensive • Slow • High capital investment	• Limited by press size • Low fiber content	• Mechanical properties modest with SMC and DMC • Material flow causes property variability • High tooling cost	• Massive tooling • Low fiber content	• One-shot process; modifications and repairs difficult • Production development usually needed on each mold

- Good acoustic and insulation properties.
- Safer handling and production in comparison with petroleum-based materials.
- No CO_2 generation.

5.4.2 DISADVANTAGES OF BIOCOMPOSITES

- Low thermal stability.
- Prone to microbial attack.
- Rate of moisture absorption is greater.
- Mechanical properties vary.
- Processing temperature is limited to 200°C.

5.5 APPLICATIONS

5.5.1 HEALTH CARE

The biocomposite material "STIMULAN" has been accepted by European countries as a carrier of antibiotics for bones and soft tissues. Biocomposite materials are also used for the reconstruction of spine, trauma, joints, foots, and ankle. Gelatin-based polymer nanocomposites are used in wound dressing. The modified cellulose textile fabric composites in-situ generated nanosilver/copper is proposed to be a better wound dressing material [13,14]. The silk fibroin-based biopolymeric composite materials possess better hydrophilicity, mechanical strength, swelling, cell attachment, and proliferation, which make the material more suitable for wound dressing. The biocompatibility of biocomposite materials makes them more suitable for biomedical applications.

5.5.2 AUTOMOBILE APPLICATIONS

The automotive parts made of biocomposite materials offer properties such as light weight, thermal resistance, and increased efficiency with low fuel consumption [15]. The E-class car interiors are made of biocomposites. Leading car manufacturers of Germany such as Mercedes, Volkswagen, Audi, and Ford utilize biocomposites for automotive spare parts, interior, and exterior parts. The car door inner trim panels are made of 60% natural fiber in a polyurethane resin [16].

5.5.3 MARINE APPLICATIONS

Biocomposite materials are found to be a better alternative for the existing synthetic fiber-reinforced composites because it possess better mechanical property and biodegradability. Flax/polylactic acid composites [17] and Kevlar-based biocomposites are inspected for their water absorption property and aging under sea water. A major setback is noted with increased water absorption, structural change, swelling at composite interface, and degradation of fibers, which reduce the mechanical properties. The abovementioned properties are enhanced by using vegetable fibers. Water absorption is reduced by providing coatings, which makes biocomposites more suitable in marine applications [18].

5.5.4 Construction Applications

The coir-polyester composites are employed in fabricating helmets, roofing and post-boxes. These components are unaffected by weathering conditions for about 6 years. Concrete panels made of biocomposites have better durability and are unaffected by acidic/basic environment. Coir-reinforced concrete panels are selected for lightweight loading structures [19].

5.5.5 Packaging Applications

Biocomposite materials with inorganic nanometal particles and vegetables/agro-wastes as fillers are used to develop packaging materials. The developed biocomposite packaging materials possess appreciable strength and shelf life and best suit packaging applications. These materials are found to be effective toward environmental factors such as humidity, temperature, and pressure [20,21]. The addition of nanometals as fillers adds antimicrobial property to the packaging material and preserves the food quality [22].

Biocomposite material finds application in various fields like textile industries in developing fire-retardant, anti-wrinkle, and stain-resistant fabrics. The material finds more applications in medical fields for the treatment of burned wounds; cellulose-based biocomposite films used for wound dressing get absorbed onto the skin.

Biocomposite materials are becoming a focused area of research with a wide variety of products with numerous applications being developed. Being environment friendly and biodegradable, biocomposite materials are considered to be the materials for the future.

REFERENCES

1. Dicker, M.P.M., Duckworth, P.F., Baker, A.B., Francois, G., Hazzard, M.K., and Weaver, P.M. (2014) Green composites: A review of material attributes and complementary applications. *Compos. Part A Appl. Sci. Manuf.*, **56**, 280–289.
2. Rana, A.K., Mandal, A., and Bandyopadhyay, S. (2003) Short jute fiber reinforced polypropylene composites: Effect of compatibiliser, impact modifier and fiber loading. *Compos. Sci. Technol.*, **63** (6), 801–806.
3. Wu, C.S., and Liao, H.T. (2014) The mechanical properties, biocompatibility and biodegradability of chestnut shell fibre and polyhydroxyalkanoate composites. *Polym. Degrad. Stab.*, **99** (1), 274–282.
4. Ferreira, D.P., Cruz, J., and Fangueiro, R. (2018) Surface modification of natural fibers in polymer composites, in Koronis, G. and Silva, A. (eds.), *Green Composites for Automotive Applications*, Woodhead Publishing Series in Composites Science and Engineering. Elsevier, Woodhead Publishing, pp. 3–41.
5. Haraguchi, K. (2014) Biocomposites, in Kobayashi S. and Müllen K. (eds.), *Encyclopedia of Polymeric Nanomaterials*. Berlin, Heidelberg: Springer, pp. 1–8.
6. Lau, K., Ho, M., Au-Yeung, C., and Cheung, H. (2010) Biocomposites: Their multifunctionality. *Int. J. Smart Nano Mater.*, **1** (1), 13–27.
7. Yıldızhan, Ş., Çalık, A., Özcanlı, M., and Serin, H. (2018) Bio-composite materials: A short review of recent trends, mechanical and chemical properties, and applications. *Eur. Mech. Sci.*, **2** (3), 83–91.

8. Reinhart, T.J. (1998) Overview of composite materials, in Peters S.T. (ed.), *Handbook of Composites*, Springer US. Boston, MA: Springer, pp. 21–33.

9. Jamir, M.R.M., Majid, M.S.A., and Khasri, A. (2018) Natural lightweight hybrid composites for aircraft structural applications, in Jawaid, M. and Thariq, M. (eds.), *Sustainable Composites for Aerospace Applications*, Woodhead Publishing Series in Composites Science and Engineering. Elsevier, Woodhead Publishing, pp. 155–170.

10. Filament Winding – Open Molding | CompositesLab. (n.d.). Available at http://composites-lab.com/composites-manufacturing-processes/open-molding/filament-winding/.

11. Sapuan, S.M., Tamrin, K.F., Nukman, Y., El-Shekeil, Y.A., Hussin, M.S.A., and Aziz, S.N.A. (2017) Natural fiber-reinforced composites: Types, development, manufacturing process, and measurement, in Hashmi, M. S. J. (ed.), *Comprehensive Materials Finishing*, vol. 1–3, Elsevier Inc., pp. 203–230. https://doi.org/10.1016/B978-0-12-803581-8.09183-9

12. Shaker K., Nawab Y., and Jabbar M. (2020) Bio-composites: Eco-friendly substitute of glass fiber composites, in Kharissova, O., Martínez, L., and Kharisov, B. (eds.), *Handbook of Nanomaterials and Nanocomposites for Energy and Environmental Applications*. Cham: Springer. doi: 10.1007/978-3-030-11155-7_108-1

13. Paramasivan, S., Nagarajan, E.R., Nagarajan, R., Anumakonda, V.R., and Hariram, N. (2018) Characterization of cotton fabric nanocomposites with in situ generated copper nanoparticles for antimicrobial applications. *Prep. Biochem. Biotechnol.*, **48** (7), 574–581. doi: 10.1080/10826068.2018.1466150.

14. Sivaranjana, P., Nagarajan, E.R., Rajini, N., Ayrilmis, N., Rajulu, A.V., and Siengchin, S. (2019) Preparation and characterization studies of modified cellulosic textile fabric composite with in situ-generated AgNPs coating. *J. Ind. Text.*, **50** (7), 1111–1126.

15. Reddy, T.R.K., Kim, H.-J., and Park, J.-W. (2016) Renewable biocomposite properties and their applications, in *Composites from Renewable and Sustainable Materials*, InTech.

16. Marsh, G. (2003) Next step for automotive materials. *Mater. Today*, **6** (4), 36–43.

17. Le Duigou, A., Bourmaud, A., Davies, P., and Baley, C. (2014) Long term immersion in natural seawater of Flax/PLA biocomposite. *Ocean Eng.*, **90**, 140–148.

18. Le Duigou, A., Davies, P., and Baley, C. (2009) Seawater ageing of flax/poly(lactic acid) biocomposites. *Polym. Degrad. Stab.*, **94** (7), 1151–1162.

19. Satyanarayana, K.G., Sukumaran, K., Mukherjee, P.S., Pavithran, C., and Pillai, S.G.K. (1990) Natural fibre-polymer composites. *Cem. Concr. Compos.*, **12** (2), 117–136.

20. Fortunati, E., Armentano, I., Zhou, Q., Iannoni, A., Saino, E., Visai, L., Berglund, L.A., and Kenny, J.M. (2012) Multifunctional bionanocomposite films of poly(lactic acid), cellulose nanocrystals and silver nanoparticles. *Carbohydr. Polym.*, **87** (2), 1596–1605.

21. Yu, H.Y., Yang, X.Y., Lu, F.F., Chen, G.Y., and Yao, J.M. (2016) Fabrication of multifunctional cellulose nanocrystals/poly(lactic acid) nanocomposites with silver nanoparticles by spraying method. *Carbohydr. Polym.*, **140**, 209–219.

22. Sivaranjana, P., Nagarajan, E., Rajini, N., Jawaid, M., and Rajulu, A.V. (2018) Formulation and characterization of in situ generated copper nanoparticles reinforced cellulose composite films for potential antimicrobial applications. *J. Macromol. Sci. Part A Pure Appl. Chem.*, **55** (1), 58–65.

6 Polymer Nanocomposites

Polymer Composites: Design, Manufacturing, and Applications

Onkar A. Deorukhkar
Dr. Vishwanath Karad MIT World Peace University

S. Radhakrishnan
Dr. Vishwanath Karad MIT World Peace University

Yashwant S. Munde
MKSSS's Cummins College of Engineering for Women

M. B. Kulkarni
Dr. Vishwanath Karad MIT World Peace University

CONTENTS

6.1 INTRODUCTION

Nanocomposites are materials of the 21st century having a yearly development rate of 25% due to their multifunctional capacities with remarkable design and properties. Polymer nanocomposites (PNs) are characterized as a blend of two or more materials, where the matrix is a polymer and the dispersed stage has at any rate one measurement lesser than 100 nm [1]. PNs comprise a polymeric material (*viz.* thermosets, elastomers, and thermoplastics) and a reinforcing nanoscale material.

Nanoparticles have at least one measurement at nanometer scale. PNs show main enhancements in gas obstruction mechanical properties, fire retardancy properties, and good strength. Nanomaterials can be grouped into nanostructured materials and nanoparticle materials. Typically, it refers to dense mass materials made of agglomerates, with grain sizes in the nanometer size range, though the last are generally dispersed nanoparticles. The nanometer size covers a wide range from 1 nm to as extensive as 100–200 nm. To recognize nanomaterials from mass, it is critical to show one of kind properties of nanomaterials and their effects in science and innovation.

The use of inorganic nanoparticles as additives into polymer frameworks has brought made PNs multifunctional, high-performance polymers compared to customary filled polymeric materials. The transformation of these new materials will empower the circumvention of characteristic material performance by attaining new properties and using distinctive synergies among materials. This is only possible when the length size of morphology and the essential material science-related properties match well. Multifunctional highlights owing to PNs comprise enhanced thermal resistance and flame resistance, charge dissipation, moisture resistance, chemical resistance, and decreased permeability. In the course of modification of the additives at the nanoscale stage, one can improve properties of chosen polymer frameworks to surpass the necessities of business, military,

and aviation applications. The specialized methodology includes the incorporation of nanoparticles into desired polymer matrix whereby nanoparticles might be
surface treated to give hydrophobic qualities and upgraded consideration into the
hydrophobic polymer matrix [2–5].

PNs can be commonly divided into three major categories based upon the
elements of the dispersed nanoscale fillers. In the first category, the two-dimensional nanoscale fillers, for example, layered silicate [6], graphene [7,8] or MXene
[9,10] as sheets of one to a couple of nanometers thick and 100–1,000 nm in length
are available in polymeric lattices. The corresponding PNs can be gathered into
the type of layered PNs. In the second type, two dimensions are in nanometer
scale and the third is bigger, forming a lengthened one-dimensional structure;
these nanoscale fillers incorporate nanofibers or nanotubes, such as carbon nanofibers and nanotubes [11] or halloysite nanotubes [12], as reinforcing nanofillers
to get materials with extraordinary properties. The third kind is nanocomposites comprising nanoscale fillers of three dimensions in the order of nanometers.
These nanoscale fillers are iso-dimensional low perspective proportion nanoparticles, for example, spherical silica [1,13], semiconductor nanoclusters [14], and
quantum dots [15].

Polymer-based nanocomposites have advanced properties (mechanical and physical) over host polymers because of the large interfacial region among nanofillers.
Circular polymer framework is considered for the evaluation of interfacial zone. A
nanomaterial broadly used to prepare nanocomposites on a business scale is nanoclay [16]. Consequently, the conversation through the rest of this assessment will take
polymer/nanoclay composites as a beginning stage, and talk about other PNs (e.g.,
CNT and nanoparticle-reinforced composites).

6.2   THREE CATEGORIES OF NANO-REINFORCEMENT

The proportion of surface area to volume of nano-reinforcements is vital to recognize the structure–property connections of PNs. Material properties depend on the
size at nanoscale, and the surface area/volume proportion is normally three times
more prominent for nano-reinforcements compared to their micron-sized parts. The
outcome is that the surface chemistry of nano-reinforcements dominates PNs properties. Nano-reinforcements might be divided into three principle categories, namely,
particles, fibers, and layered materials.

6.2.1   NANOPARTICLES

To increase elastic modulus and yield quality, in ordinary nanofilled composites,
micron-sized particles are commonly added as reinforcements.

On the other hand, scaling these molecule reinforcements down to nanometer
scale can give significant novel material properties, for example, gold nanoparticles
are red in color and silver nanoparticles transform color contingent upon their shape
and size.

6.2.2 LAYERED PLATELET MATERIALS

There are three kinds of layered platelet nanomaterials of industrial interest for polymer-filled nanocomposites, including naturally available and synthetic clays like montmorillonite (MMT), saponite, mica, graphite, and graphene. Among these, layered MMT clays comprising nanometer thick platelets are the most financially efficient nanomaterials, and are economically feasible. MMT is the essential part of bentonite mineral clay, which additionally contains minerals, for example, quartz, mica, feldspar, and zeolite; bentonite is acquired either by quarrying or mining based on its resource [17]. Nanoclay nanocomposites provide enhanced strength, quality, durability, and thermal stability, as well as gas permeability and lower thermal coefficient [18,19].

Nanoclays are bulk form of layered materials [20]; these layers can be separated and scattered in the nanofilled matrix to form a nanocomposite material.

A few methodologies have been developed to prepare PNs utilizing nanoclays based on the polymer matrix employed. Naturally occurring nanoclay chemical structures.

Polymer reinforcements utilizing fillers, either inorganic or organic, are regular in new plastics advancements. Polymer nanostructured materials (PNMs) are a radical option in contrast to these conventional filled polymers or polymer pieces. As opposed to conventional polymer matrices where reinforcement is on the order of microns, PNs are exemplified by discrete components on the order of a couple of nanometers. Uniform scattering of these nano-sized fillers produces ultra-large interfacial area per volume among the nanoparticles and the host polymer. Enormous interfacial region and nanoscopic dimensions among nanoparticles generally separate polymer nanostructured composites from customary filled plastics and composites. These material qualities infer that the general performance of PNMs cannot be comprehended by easy scaling systems that relate to conventional polymer composites. Along these lines, novel blends of characteristics obtained from the nanoscale structure of PNMs give chances to go around customary performance related with traditional reinforced plastics, typifying the assurance of PNMs [2].

The assessment of the nanofiller scattering in the nanofilled matrix is significant because the mechanical and thermal properties are identified with the acquired morphologies. Based on the degree of separation of the nanoparticles, three kinds of nanocomposite morphologies are feasible, including microcomposites, intercalated nanocomposites, and exfoliated nanocomposites. At the point when the polymer cannot intercalate between silicate layers, a composite of different phases is attained, whose characteristics are in a similar range as those viewed in conventional composites [21]. Intercalated structure, wherein a solitary broadened polymer chain is intercalated among the layers of the silicate, brings about an all-around ordered multilayer morphology with intercalated layers of polymer and earth. At the point when the silicate layers are completely and uniformly scattered in a continuous polymer matrix, an exfoliated structure is obtained [21]. Furthermore, in this investigation, different sorts, handling strategies, and various applications are discussed thoroughly.

6.3 NANOCOMPOSITE TYPES

There are various sorts of economically available nanoparticles that can be included into the polymer matrix to form PNs. Depending upon the application, the researcher must decide the kind of nanoparticles expected to give the ideal impact. A concise elaboration incorporates the generally utilized nanoparticles in the literature.

6.3.1 POLYMER-BASED MATRICES

6.3.1.1 Thermoplastic Polymers

Thermoplastic polymer material with direct chain atoms is softened by exposure to heat and solidification using cooling at different temperatures [22]. Thermoplastic polymers are fit to be shaped for required structure because on their softened impressionable state. Additionally, high-molecular-weight thermoplastic polymers have various sorts of bonds, for example, weak van der Waals forces (polyethylene), strong dipole-dipole interactions, hydrogen bonding (nylon), and aromatic rings (polystyrene) [22].

These polymers are separated in two categories, namely, amorphous and glasslike based on changing temperatures. Different characteristic is quick modulus reduction that happens in amorphous thermoplastics over the glass transition temperature fluid state. The liquefying temperature of crystalline stage and glass transition temperature of coinciding amorphous stage are assumed as the basis for preparing thermoplastic polymers. After cooling, step crystallization happens quickly [23].

The crystallinity level of thermoplastic polymers relies upon the cooling time. Additionally, either elevated level temperature or long stay time at a particular temperature may change the polymer qualities, particularly mechanical properties. Commonly, in composite frameworks, there are three principle matrix types, namely, polymer, metal, and clay, with dissimilar added substances in different forms, such as fillers, lamina, piece, and fibers.

6.3.1.2 Thermosetting Polymer Matrices

Thermosetting resins are less viscous fluids or low-molecular-weight solids that require additives such as cross-connecting agents to be formulated and treated. Likewise, these polymers can be doped with fillers to enhance thermal and mechanical properties [23]. Among the curing procedure, which utilizes heat and pressure, thermoset resins completely polymerize and slowly solidify with polymerization process and cross-connection of the molecules. These polymers show different attributes through their three-dimensional cross-connected structure with great resistance to solvents, high stability, and resistance to high temperature [24]. Thermoset molecules can react uninhibitedly to give covalent bonds; in this way, cross-connecting is achieved in one massive particle among the polymer as an exothermal reaction. This is because of the impact of molecules to a lower energy state than the random molecule orientation of the fluid. Thus, these particles are bound together by means of covalent bonds and cannot be softened through warming [25,26]. Thermosetting resins incorporate various kinds of polymers, such as bismaleimides, epoxies, vinyl esters, polyamides, and polyesters [27]. Several researchers have utilized thermoset frameworks in polymer composite frameworks, particularly PNCs. Gorowara et al. [28] considered the molecular aspects of glass fiber surface coatings in thermosetting composite frameworks containing polymer matrix. In this work, multicomponent glass fiber was explored for coating bundles utilized in business glass fiber make. Kim et al. [29] examined the impacts of the size of silver pieces and their distribution on the thermal and electric conductivities of a polymer-based composite. Another investigation [30] employed a molecular dynamics (MD) simulation method to break down the fragile crack in epoxy-based thermoset polymer with mechanical stacking.

A researcher studied that the ductile appearances of amorphous polymers were concentrated by conventional MD simulation approach by means of stress–strain response to yield point [31]. Several strategies are available in the literature that can be employed to reuse thermosetting polymer. Thermosetting polymers cannot be remolded because of their cross-connected position. However, various thermosetting polymers, for example, polyurethane, can be changed effectively to their initial monomer, despite the fact that the more predominant thermosetting resins (epoxy and polyester) cannot be depolymerized to their main components [31].

6.3.1.3 Polymer Matrices and their Applicability

6.3.1.3.1 Polyamide Matrices

In polyamide polymers, amide groups are the key components [32,33]. Polyamide polymers represent some unique abilities, such as, great stability of dimensions, mechanical quality, high heat temperature, great resistance to chemicals and oxygen, vibration, along with insulation for electrical appliances [32]. One of the uses of polyamide polymers is in flexible films that are appropriate for food packaging [32].

Ecologically, biopolyamides are characterized as polyamides that are mostly derived from renewable sources [34].

6.3.1.3.2 Polypropylene and Polyethylene Matrices

Polypropylene (PP) is a type of thermoplastic that has been developed or its unique qualities (flexible physical properties, minimal effort and thickness, and high heat distortion temperature (HDT)) [35]. Characteristics are highlighted to the polymer chains and the monomers [36]. For the most part, PP contains three primary sorts with various characteristics comprising PP blocks copolymers, homopolymers, and PP random copolymers. PP is utilized for different applications (filaments, spunfortified nonwovens, and tapes) [36]. Generally, thermal, chemical, biological, and mechanical are recognized as degradation methods. Polymer degradation is a typical issue that has been investigated previously. The volatile component loss of mechanical properties has been illustrated [36]. In some studies, polyethylene (PE) was blended with PP to upgrade physical properties, such as removal of expensive synthesis for new block copolymers and low-temperature performance [37]. PE is also used in food packaging and containers. This is because of resistance of this polymer against biodegradation, enzymatic degradation that causes a few significant ecological dangers, named "white pollution" [38]. Kamrannejada et al. [38] explored the photocatalytic degradation of carbon-covered TiO_2 nanoparticles in PE-based nanocomposites.

6.3.1.3.3 Liquid Crystal Matrices

Liquid crystal polymers (LCPs), as a group of thermoplastics, contain a highly crystalline molecular chain as opposed to other common polymers (e.g., nylon). Hence, this property makes these polymers almost linear, semi-rigid, stacked orientation of molecules with profoundly ordered status in the liquid crystal stage [39,40]. Anisotropy characteristic of LCPs and additional capability to separate are because of the bonds inside the particle that result in a high force inside the molecule. In this issue, if force is transitionally applied to the particle orientation, the minor bonds get a large portion of the mass that causes a simpler separation. In contrast, a longitudinal force intensely influences the essential bonds of the particles and results in an elevated separation [39,40].

6.3.1.3.4 Polyurethane Matrices

Primarily, polyurethane is connected with chemicals as urethane polymers, which contain two main raw materials (polyols and isocyanates). These materials are combined using catalysts and several additives. Based on the kinds of isocyanates and polyol constituents, the last product status is determined and can be explained as froth and solid/fluid. PU composites are characterized with regards to the fiber reinforcement and compactness of polymeric matrix. Khudyakov et al. [41] revealed PU nanocomposites of photopolymerizable (UV-reparable) by mixing nanosilica and organoclay. Along with the uses of PUs, despite the fact that the solution-based PU is mainly utilized in various industries, water-based PU resin has attracted interest for its ecologically well-disposed attributes and fire safety [42]. Ibrahim et al. [43] blended a renewable source polymer as a host in polymer electrolyte for electrochemical gadgets. In this study, polyurethane and LiI and NaI were blended in various wt% to manufacture a film of polymer electrolytes.

6.3.1.3.5 Epoxy Matrices

A resin associated with epoxy was presented by Pierre Castan in 1938. Nowadays, epoxy resins are generally utilized as protecting coatings or auxiliary applications (*viz.*, composites, cements, and projecting) [44,45]. Among thermosets, the epoxy resin can show the best mechanical properties and ecological resistance along with a few advantages. Moreover, these resins can be used at room temperature to cover the necessary space as a result of their low viscosity. Likewise, they provide low shrinkage that leads to minimum residue stress after cross-linkage. In PNC frameworks, epoxy resins are reinforced by utilizing nanofillers to obtain better properties, such as chemical and mechanical properties.

Typically, the epoxy resins have more than one epoxide bunches for each particle in their structure. Industrial epoxy resins are aliphatic or aromatic spines, cycloaliphatic, which are produced using either epichlorohydrin or direct epoxidation of olefins and peracids. Generally, they are used as an intermediate for epoxy resins. Diglycidyl ether of bisphenol is manufactured using bisphenol A and excess epichlorohydrin.

6.3.1.4 Polymer-nonmetallic Nanocomposites

6.3.1.4.1 *Polymer-carbon Nanotubes*

Since 1991, carbon nanotubes (CNT) have been used. While Iijima presented it as modern developing material with surprising qualities (mechanical and electric). Usually, there are two categories of CNT comprising single-walled nanotubes (SWNT) that have a single graphite sheet flawlessly wrapped into a tube-shaped cylinder and multiwalled nanotubes (MWNT) with a group of nanotubes [46,47]. Polymer nanocomposite has fascinating applications as CNTs. Doping ameliorates mechanical, electric, and thermal factors of polymer networks. In particular, the mechanical properties of CNT, particularly high rigidity and sturdiness, accomplish more applications in the manufacturing of PNCs. Furthermore, high aspect ratio allows them to be adjusted along one axis and makes them usable in production of conductive polymers for accomplishing the necessary electric properties [47].

6.3.1.4.2 *Graphene-Based Polymers*

It is a hexagonal sp2 single carbon atomic two-dimensional material sheet with incredible mechanical, transport, and thermal properties. In particular, depending upon its excellent mechanical properties, it may be presented as strengthening components of composites [48–58]. Graphene, which is used for the production of graphene-based PNCs, is generally produced by chemical or thermal treatment of graphite oxide. This procedure has more merits because of the scale at which graphene can be formed. Mainly, three major procedures are utilized to manufacture the graphene/PNCs comprising in-situ polymerization, processing of solvent, and melt processing method [59]. Some researchers have surveyed PNCs on graphene-based fillers and presented the advancement in processing, capacities, and performances of graphene-polymer composites [60,61].

6.3.1.4.3 *Polymer Nanodiamond*

Nanodiamond (ND) exhibits numerous vital characteristics of bulk diamond, for example, hardness, Young's modulus and thermal conductivity, that make it a valuable material for various uses. Based on these significant qualities of NDs, they are utilized in numerous usage, like thermal conducting, electronic and optic mechanics, heat interface, greasing up oil, biomedical innovation, and fillers for nanocomposite networks [62,63].

6.3.1.4.4 *Clay-based Polymer*

Nanoclays are presented as ecologically attractive materials with minimal cost. In this polymer, grades, characterization, synthesis, surface element, stability, and production of polymer/nanoclay are explored for building up novel materials.

In polymer/nanoclay composites, the ultimate qualities are transformed rather than bulk polymer because of molecule reinforcing and more confined development in polymer chain as a result of cross-linking network. In this area, the intercalated nanoclay layered with polymers outlines the most anticipated essential factors from polymer or nanoclay composites. There are different strategies that are applied so as to add nanoclays among polymer frameworks, such as chemical and mechanical [64,65]. Zabihi et al. [66] explored MMT nanoclay in self-assembly of quaternized chitosan nanoparticles inside mud layers. Sari et al. [67] applied surface change of nanoclay particles by a few measures of polyester-amide hyperbranched polymer.

6.4 NANOCOMPOSITES PROCESSING

Numerous methods are used for the processing of the PNs. The most significant techniques are discussed in this section.

6.4.1 INTERCALATION OF THE POLYMER

Intercalation method is utilized for the production of polymer-based nanocomposites, which contain layered silicates. In this technique, a solvent is employed in which the polymer or prepolymer is soluble and the silicate layers are swellable. Nanocomposites produced using this technique have structures extending from intercalated to exfoliated based on the penetration force of the polymer chains into the silicate materials. Thus, this has become a standard technique for the production of polymer-layered silicate nanocomposites [68].

At the point when the polymer cannot intercalate among the silicate sheets, a phase-separated composite is achieved, which has similar properties as that of conventional microcomposites. Once polymer matrix enters the layered silicates, intercalated nanocomposites structure is obtained in a crystallographically normal manner, despite the clay to polymer proportion. An ordered multilayer morphology developed with substituting polymeric and inorganic layers is produced. Normally, just a couple of molecular layers of polymer may be added in these materials [68].

6.4.2 DIRECT MIXING

Entrenched polymer handling strategies are utilized for direct blending. For instance, nanocomposites with PP and calcium carbonate ($CaCO_3$) matrix were set up through melt blending of the constituents via Haake blender [69]. $CaCO_3$ nanoparticles are 44 nm in width. This strategy creates scattered nanocomposite samples at generally less filler volume of 4.8% and 9.2%, yet accumulation happens at a moderately high of 12%. A form of soften blending was similarly used to produce nanocomposites. Silica-polyurethane nanocomposites were prepared by first blending silica with polyol [70]. After that, the blend was cured with diisocyanate at 100°C for 16 h with catalyst (0.1%). The circular particles have a normal width of 12 nm and a narrow size distribution of 10–20 nm. Excellent quality samples were prepared at several wt% filler fractions. High-quality dispersion of nanoparticles in polymer frameworks is one of the major issues in preparing good characteristic polymer samples. Rong et al. explained that fixing of monomers made of styrene to encompass nanoparticles yields great dispersion. Isotactic PP is utilized as the matrix and silica of a size of 7 nm is used as nanofiller [71].

6.4.3 IN-SITU POLYMERIZATION

Another method that is usually incorporated for the production of PNs is in-situ polymerization [72–76]. In this strategy, nanoparticles are first diffused in monomer and the resultant blend is polymerized. Yang et al. [73] studied that silica/polyamide-6 nanocomposites are prepared by first drying the silica particles to remove water absorbed on their surfaces. Afterwards, the particles are mixed with ε-caproamide, and an appropriate polymerization initiator is added simultaneously. Then, the blend is polymerized at a high temperature under N_2 [73]. This method offers dispersed nanocomposite samples with silica of ~50 nm; however, aggregates are checked for small particles (~12 nm) [74].

The process for the production of PNs through this method is shown in Figure 6.1. This is due to the improved surface energy for small size particles, which helps further particle aggregation.

6.4.4 SOLUTION MIXING

Solution mixing is used to prepare particulate PNs. This method is also utilized to prepare polymethylmethacrylate (PMMA) nanocomposites with good dispersion of alumina nanoparticles [77,78]. Alumina nanoparticles are initially added to methylmethacrylate (MMA) monomer and dispersed in a less viscous solution through sonication. Later, an initiator and chain transfer agent is included. Then, the blend is polymerized under N_2 and dried in vacuum. Next, processing to make samples is done by traditional methods.

6.4.5 PROCESSING OF IN-SITU PARTICLES

This strategy, specifically in-situ sol-gel handling of nanoparticles inside polymers, permits structure of organic–inorganic nanocomposites. This procedure is utilized for preparing particulate PNs with silica and titania in different polymer matrices [79–87]. In this system, a few different modes can be used to create PNs. The first is to blend silica precursor with a copolymer of polymer matrix, and then the reaction of sol-gel proceeds to create silica. Through drying, the polymer blocks separate phases and the silica areas combine [80]. The next step is to blend a silica precursor,

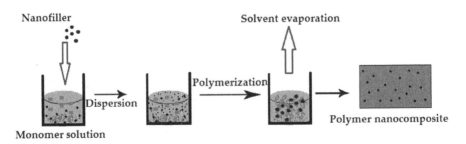

FIGURE 6.1 Diagrammatic representation of in-situ polymerization technique.

for example, tetraethylorthosilicate (TEOS), with a polymer matrix (polyimide and epoxy) [85,88].

This method has additionally been used for the production of TiO_2/PNs. Direct addition of TiO_2 nanoparticles to polystyrene maleic anhydride (PSMA) results in aggregation. Accordingly, PSMA is first dissolved in tetrahydrofuran (THF), and afterward tetrabutyl titanate is added under proper parameters to produce nano-composites with dispersion of TiO_2 nanoparticles [86]. The uncondensed TiOH and maleic anhydride can be reacted and the coating of polymer on the titania nanoparticles can be applied to avoid agglomeration. Likewise, dispersion of TiO_2 in PSMA can be promptly accomplished. Nanocomposite films (Nano-TiO2/poly-imide) are acquired utilizing the sol-gel process [87,89]. Two different ways have been utilized for production of PNs, for example, silica/polydimethylsiloxane (PDMS) composites [90]. PDMS matrix is first produced with TEOS as an end-connecting operator. At that point, the system is swollen with TEOS and the sol-gel response is catalyzed.

6.4.6 WET CHEMICAL PROCESSING

In this type of processing, suspensions are utilized for the formation of thin layer films, coming about in casting or coatings on various substrates. Polymer-based coat-ings are applied to a substrate by means of lacquering, which are the principally used procedures in processing [91]. Drying methods shift from ambient conditions drying to traditional hot air drying and infrared to microwave energy drying, with every technique impacting the film or coating characteristics [92]. The applicable coat-ing weight can be very low, keeping up the ideal barrier properties compared with extrude coatings [93].

6.4.7 THERMOPLASTIC PROCESSING

Thermoplastic production of polymers is mostly executed by means of extrusion and is the most essential polymer processing strategy. Extrusion permits melting a poly-mer with a high energy input in less time. Because of the supply of heat and energy brought about by friction between the screws, the mass melts, gets formable, and is pressed through the extruder die [94]. During the entire procedure, the mass can be packed, blended, homogenized, chemically transformed, and plasticized [95,96]. On including nanoparticles into polymeric compounds, various kinds of nanocomposites are feasible. Nanocomposites are processed using the dispersion characteristic based on the extruder and screw arrangement [97].

6.4.8 COMMON PROCESSING METHODS

Most well-known methods for the deposition of thin layers belong to the type of gas-phase routes. These procedures permit to coat flexible substrates with nanoscale polymeric layers with thicknesses in the nanometer scale up to about several nanometers [98]. Depending upon the deposition mechanism, gas-phase procedures are categorized as physical or chemical vapor deposition (PVD/CVD).

PVD methods are performed in a high vacuum and depend on the exchange of the solid coating material into the gas or vapor phase followed by the condensation on top of the substrate [99]. Characteristic materials to be deposited are metals and metalloids just as their nitrides, carbides, and oxides. Deposition of compounds is conceivable by reactive process, which depends on a chemical reaction between the material and gases such as O_2 and N_2. The vapor phase of the material coating is acquired by evaporation because of electrical resistance heating or electron ray irradiation [99]. It permits the deposition of barrier layers which are appropriate for the encapsulation of organic electronic gadgets. The huge quantity of layers prepared by CVD or PVD methods on adaptable substrates is employed in food packaging technology [99].

In the next procedure, atoms of the strong coating material are released because of the effect of particles of the process gas which are ionized via an electrical field. The distinctive and barrier characteristics of the deposited layer can be enhanced by applying a magnetic field during the procedure [100]. Fabrications of thin metal films on polymeric surfaces using PVD are normally utilized methods for the class of metal-PNs [101–105]. CVD methods depend on chemical reactions on the substrate [99,100]. The reactions are actuated either by heat or by plasma. These procedures permit the deposition of inorganic or polymeric materials such as grapheme, SiOxCyHz, and parylene [102,103,105]. Atomic layer deposition is customized version of a CVD procedure.

6.5 ADVANTAGES AND DISADVANTAGES

PNs have a few advantages that are listed below [104]:

1. They are lighter than traditional composites as high degrees of solidness and quality are acknowledged with far less high-thickness material.
2. Barrier properties are improved compared with the perfect polymer.
3. Mechanical and thermal properties are possibly superior.
4. Excellent flammability characteristics and enhanced biodegradability of biodegradable polymers are possible.

In the progress of PNs, there are a few difficulties and constraints.

For instance, polymeric nanocomposites require controlled blending/intensifying, stabilization of the dispersion, and orientation of the dispersion stage. Despite the fact that the modulus of polymeric nanocomposites increases with increasing nanofiller substance and sturdiness, it affects the quality which diminishes when the materials become more fragile [105,106]. With the nanofiller content, the viscosities of polymeric nanocomposites increases, which make production difficult [107]. Additionally, highly viscous flow of polymer melts brings huge forces for the duration of injection molding and extrusion. Especially, the impact of nanofiller on the polymer characteristics are different from anticipated via the thermodynamic studies for reduced size of particle filler [108]. Some other advantages and limitations are also reported in the literature which are listed in Table 6.1 [109].

TABLE 6.1

Advantages and Limitations of Polymer-based Nanocomposite Processing Methods [109]

Process	Advantages	Limitations
Intercalation/ prepolymer from solution	Synthesis of intercalated nanocomposites based on polymers with low or even no polarity. Preparation of homogeneous dispersions of the filler	Industrial use of large amounts of solvents
In-situ intercalative polymerization	Easy procedure based on the dispersion of the filler in the polymer precursors	Difficult control of polymerization. Limited applications
Melt intercalation	Environmentally benign; use of polymers not suited for other processes; compatible with industrial polymer processes	Limited applications to polyolefins, which represent the majority of used polymers
Template synthesis	Large-scale production; easy procedure	Limited applications; based mainly in water-soluble polymers, contaminated by side products
Sol-gel process	Simple, low processing temperature; versatile; high chemical homogeneity; rigorous stoichiometry control; high purity products; formation of three dimensional polymers containing metal-oxygen bonds. Single or multiple matrices. Applicable specifically for the production of composite materials with liquids or with viscous fluids that are derived from alkoxides	Greater shrinkage and lower amount of voids compared to the mixing method

6.6   APPLICATIONS

Numerous applications of PNs and several industrial products are accessible in the market. Remarkable areas are automotive parts [110,111], packaging [112], materials construction [113], biotechnology [114], clinical devices, and so on. One significant characteristic seen because of the incorporation of clay particles in polymers is flame retardancy. This quality gives avenues to applications in numerous areas, for example, building materials, computer parts, and vehicle interiors [113]. Another area of application is in industries related to packaging.

The impermeable clay layers mandate an indirect path, which make the dispersion of particles all through the matrix challenging. There are many applications with graphene and CNTs. A region of important application of these materials is in the optoelectronic industry [115,116]. The utilization of a aqueous diffusion of PE copolymer with a moderately high substance of acrylic acid and as an optional medium to get PE NFC nanocomposites involved ongoing investigation [117]. Water suspensions of NFC with xylan, pectin, and xyloglucan were investigated for frothing and basic properties as other methods for structuring of food [118]. Some applications are discussed in this section.

6.6.1   GAS BARRIERS PROPERTIES (FOR PLASTIC BOTTLES, PACKAGING, AND SPORTS GOODS)

Mixtures made of polydimethyl siloxane rubber and nanosilica formed in-situ by hydrolysis of tetraethyl orthosilicate may be explicitly molded, producing items, for example, golf balls. Various PNs dependent on polymers, like styrene butadiene rubber, butyl rubber, ethylene vinyl acetate copolymer, ethylene propylene diene monomer rubber, and ethylene-octene copolymer, have been utilized industrially because of their barrier properties [119,120]. These polymers can act as excellent barriers for some gases, for instance, N_2, CO_2, O_2, and chemicals, viz., toluene, H_2SO_4, HCl, HNO_3, and so forth. Likewise, PNCs have been utilized in food packaging and plastic compartments, both rigid and flexible. Explicit cases contain packaging for cheese, meat, cereals and dairy items, printer cartridge seals, clinical holder seals for blood collecting tubes, stoppers for clinical containers and blood bags, child pacifiers, and drinking water bottles. The incorporation of layered silicate clay diminishes dispersion rate of smaller particles, for example, water and oxygen, polymer films, and so on [121].

Dual core Wilson tennis ball is one of the newly commercialized sports products. The clay nanocomposite covering of the Wilson tennis balls maintains the inner pressure for a longer time. The core is covered by a butyl rubber-clay nanocomposite that is used as a gas barrier, doubling-up its shelf life. Using PNs more adaptable coatings with gas permeability of up to 30 to 300 times lower than butyl rubber have been formed. These coatings have been demonstrated to be intact by strains up to 20%. This dual core new tennis balls utilizing this coating maintain air longer and can bounce twice as high compared to regular balls [122].

6.6.2   ENERGY STORAGE SYSTEMS AND SENSORS

Fuel cell plays an important role in electrochemical devices, which convert chemical energy of C, H_2, and O_2 directly and proficiently into valuable electrical energy with water and heat as the main side products. Because of incorporation of nanomaterials, their productivity enhances significantly. In fuel cells, proton exchange membrane's functions are to permit proton transport from the anode to the cathode, to be an electron nonconductive material, and as a barrier for separation [123]. Common membranes are made of acidic organic polymers, like, sulfonic and carboxylic groups, which separate when solvated with water, permitting H_3O^+ (hydrated proton) transport. Sulphonated polystyrene ethylene butylene polystyrene nanocomposites can be utilized as proton exchange membranes because of their superior proton exchange ability. A number of polymers are being utilized in fuel cells applications, for example, hyperbranched polymer with a hydroxyl group at the margin, cross-linked sulphonated polyether ketone, phosphoric acid-doped polybenzimidazole, sulfonated polyarylenethioethersulfone, sulfonated polybenzimidazole copolymer, sulfonated polybenzimidazole, and so on. Clay built-in elastomers are being used as sensors to recognize fatigue, impact, and huge strain for aerospace applications.

6.6.3   MEMBRANES AND OPTICAL GLASS

Clay-based polymers are widely used for coating of transparent materials to increase both hardness and toughness of the materials without disturbing the transmission properties of light. A capacity to oppose high speed impact with considerably improved abrasion resistance of PLS nanocomposites was shown by Triton Systems. Due to this and improved optical properties, it has been broadly popularized in contact lens and optical glass usages [122].

6.6.4   ELECTRONICS AND AUTOMOBILE SECTORS

Several applications in electronic and automobile sectors of thermoset/clay nanocomposites are reported in the literature. The capacity of nanoclay incorporation to diminish solvent transmission through polymers, like polyimides, specialty elastomers, poly urethane, and so on, has been illustrated by researchers. Investigations reveal the important decreases in fuel transmission through polyamide–6/66 by incorporation of nanoclay reinforcement. The major driving force for the utilization of PNCs by tire industries is decrease in weight and manufacturing costs. Clay is one of the easily accessible materials, has extremely low density, and encourages decrease in weight. Clay built-in tires show phenomenal mechanical properties instead of using ordinary tires just as improved gas barrier performance for tubeless tire usage. In general, for automobile tire manufacturing, styrene butadiene and natural rubber-based clay nanocomposites are mostly used, and butyl rubber is used in tubes [124].

6.6.5   COATING MATERIALS

For adjusting properties of surfaces, coatings are important. A few methodologies have been attempted by scientists for improving surface properties of coatings. One of the improvements is nanoclay-based polymer coatings [125,126]. Nanoclay-incorporated thermoset polymer nanocoatings show excellent properties, like superhydrophobicity, improved wettability, great resistance for chemicals, corrosion resistance, improved climate resistance, enhanced abrasion resistance, superior barrier properties and impact resistance, scratch, and so forth.

6.6.6   BIOMEDICAL SECTORS

In the biomedical sector, the adaptability of the nanocomposites is favorable, which permits their utilization in a extensive scope of biomedical applications as they fulfill numerous applications in clinical materials, for instance, biodegradability, biocompatibility, and mechanical properties [127]. For this cause and because they are finely modulated by including various clay contents, they can be useful in tissue

TABLE 6.2

Potential Applications of Polymer Nanocomposite Systems [130]

Nanocomposites	Applications
Polycaprolactone/SiO$_2$	Bone-bioerodible for skeletal tissue repair
Polyimide/SiO$_2$	Microelectronics
PMMA/SiO$_2$	Dental application, optical devices
Polyethylacrylate/SiO$_2$	Catalysis support, stationary phase for chromatography
Poly(p-phenylene vinylene)/SiO$_2$	Non-liner optical material for optical waveguides
Poly(amide-imide) / TiO$_2$	Composite membranes for gas separation applications
poly(3,4-ethylene-dioxythiophene)/V$_2$O$_5$	Cathode materials for rechargeable lithium batteries
Polycarbonate/SiO$_2$	Abrasion-resistant coating
Shape memory polymers/SiC	Medical devices for gripping or releasing therapeutics within blood vessels
Nylon-6/LS	Automotive timing belt – TOYOTA
PEO/LS	Airplane interiors, fuel tanks, components in electrical and electronic parts, brakes and tires
PLA/LS	Lithium battery development
PET/clay	Food packaging applications. Specific examples include packaging for processed meats, cheese, confectionery, cereals and boil-in-the-bag foods, fruit juice and dairy products, beer and carbonated drinks bottles
Thermoplastic olefin/clay	Beverage container applications
Polyimide/clay	Automotive step assists – GM Safari and Astra Vans
Epoxy/MMT	Materials for electronics
SPEEK/laponite	Direct methanol fuel cells

engineering: the hydrogel form, in dental applications, in bone replacement and fix, and in medication control discharge.

Stable scattered CNTs by biopolymer set up into biomedical uses, tissue engineering, and drug delivery method [128]. For the bioactivity of biopolymer, their composites with CNTs give significant sensing performance. The biomimetic actuation established on CNT-filled biopolymer devices have huge and quick actuation dislocation under small voltage electrical stimulation [111,129]. Several other applications are shown in Table 6.2.

6.7   CONCLUSIONS

PNs offer incredible opportunities to investigate new functionalities compared to customary materials. The field of nanocomposites has been one of the most encouraging and advancing research areas. They are of special interest due to their unique characteristics, for example, light weight, ease of production, and flexibility. A defining characteristic of PNs is that the small size of the fillers leads to a large increase in interfacial area compared to conventional composites. The interfacial area creates a remarkable volume portion of interfacial polymer with qualities dissimilar from the bulk polymer even at small loadings of the nanofiller. Interfacial structure is known to be different from bulk structure, and in polymers with nanoparticles possessing high surface areas, most part of the polymers are present near the interfaces, despite the small weight of the filler. This is one of the reasons why the nature of reinforcement is different in nanocomposites. The properties of composites such as filler size, shape and aspect proportion, and filler network interactions are significant parameters which determine the effect of fillers.

Among numerous profoundly hyped technological items, PNs have lived up to the expectation. PNs exhibit superior properties, for example, mechanical, barrier, optical, etc., compared to smaller-scale or macrocomposites. Therefore, PNs can be seen in different fields of application. PNs for different applications can be synthesized by proper selection of lattice or matrix, nano-reinforcement, synthesis method, and surface change of either the reinforcement or polymer. Numerous products based on PNs have been commercialized. This chapter has attempted to emphasize different types of nanocomposites, processing techniques with some unique applications, and advantages of nanocomposites in different technological applications with some specific examples of industrial products.

REFERENCES

1. K. Müller, E. Bugnicourt, M. Latorre, M. Jorda, Y. S. Echegoyen and J. Lagaron, "Review on the processing and properties of polymer nanocomposites and nanocoatings and their applications in the packaging, automotive and solar energy fields," *Nanomaterials*, 7, 74–121, **2017**.
2. Joseph H. Koo, *Polymer Nanocomposites, Processing, Characterization, and Applications*, McGraw-Hill Nanoscience and Technology Series, Editor: Omar Manasreh, doi:10.1036/0071458212, McGraw-Hill, New York, **2006**.
3. G. Cao, *Nanostructures and Nanomaterials: Synthesis, Properties & Applications*, Imperial College Press, London, **2004**.
4. R. Krishnamoorti and R. A. Vaia, "Polymer nanocomposites: Synthesis, characterization, and modelling," ACS Symposium Series 804, ACS, Washington, DC, **2001**.
5. T. J. Pinnavaia and G. W. Beall, *Polymer-Clay Nanocomposites*, John Wiley & Sons, New York, **2000**.
6. M. T. Albdiry, B. F. Yousif, H. Ku and K. T. Lau, "A critical review on the manufacturing processes in relation to the properties of nanoclay/polymer composites," *J. Compos. Mater.*, 47, 1093–1115, **2013**.
7. R. J. Young, I. A. Kinloch, L. Gong and K. S. Novoselov, "The mechanics of grapheme nanocomposites: A review," *Compos. Sci. Technol.*, 72, 1459–1476, **2012**.
8. J.H. Du and H.M. Cheng, "The fabrication, properties, and uses of graphene/polymer composites," *Macromol. Chem. Phys.*, 213, 1060–1077, **2012**.
9. X. Q. Li, C. Y. Wang, Y. Cao and G. X. Wang, "Functional MXene materials: Progress of their applications," *Chem. Asian J.*, 13, 2742–2757, **2018**.
10. S. B. Tu, Q. Jiang, X. X. Zhang and H. N. Alshareef, "Large dielectric constant enhancement in MXene percolative polymer composites," *ACS Nano*, 12, 3369–3377, **2018**.
11. P. Calvert, "Nanotube composites – A recipe for strength," *Nature*, 399, 210–211, **1999**.
12. M. X. Liu, Z. X. Jia, D. M. Jia and C. R. Zhou, "Recent advance in research on halloysite nanotubes-polymer nanocomposite," *Prog. Polym. Sci.*, 39, 1498–1525, **2014**.
13. E. Reynaud, C. Gauthier and J. Perez, "Nanophases in polymers," *Rev. Metall.*, 96, 169–176, **1999**.

14. N. Herron and D. L. Thorn, "Nanoparticles: Uses and relationships to molecular cluster compounds," *Adv. Mater.*, 10, 1173–1184, **1998**.

15. P. Huang, H. Q. Shi, S. Y. Fu, H. M. Xiao, N. Hu and Y. Q. Li, "Greatly decreased red-shift and largely enhanced refractive index of mono-dispersed ZnO-QD/silicone nano-composites," *J. Mater. Chem.*, 4, 8663–8669, **2016**.

16. F. Uddin, "Clays, nanoclays, and montmorillonite minerals," *Metall. Mater. Trans. A*, 39, 2804–2814, **2008**.

17. Kunimine Industries, Bentonite (http://kunimine.co.jp/english/bent/bent_01.htm#Q7), **2015**.

18. R. S. Sinha and M, Okamoto, "Polymer/layered silicate nanocomposites: A review from preparation to processing," *Prog. Polym. Sci.*, 28, 1539–1641, **2003**.

19. D. R. Paul and L. M. Robeson, "Polymer nanotechnology: Nanocomposites," *Polymer*, 49, 3187–3204, **2008**.

20. L. Smart and E. Moore, *Zeolites and related structures Solid State Chemistry—An Introduction*, 2nd ed., Chapman and Hall, London, 238–272, **1995**.

21. M. R. De Moura, M. V. Lorevice, L. H. C. Mattoso and V. Zucolotto, "Highly stable, edible cellulose film incorporating chitosan nanoparticles," *J. Food Sci.*, 76, N25–N29, **2011**.

22. M. Mondal, "Polypropylene and natural rubber based thermoplastic vulcani-zates by electron induced reactive processing," [A dissertation submitted to the FakultatMaschinenwesen, Institut fur Werkstoffwissenschaft]. Germany: TechnischeUniversitat Dresden, **2013**.

23. M. Xanthos, *Polymers and Polymer Composites, Part One: Polymers and Fillers*, WILEY-VCH Verlag GmbH & Co. KGaA, Weinheim, **2010**.

24. Congress of the U.S. Office of Technology Assessment, "Advanced materials by design, polymer matrix composites," Building materials 22(46), DIANE Publishing, ISBN:9781428922396, **1988**.

25. A. Tomas and B. Strom, *Manufacturing of Polymer-matrix Composites*, CRC Press, Boca Raton, FL, 1997.

26. D. Feldman, "Some considerations on thermosetting polymers as matrices for compos-ites," *Prog. Polym. Sci.*, 15, 603–628, **1990**.

27. E. Klata, S. Borysiak, K. Van de Velde, J. Garbarczyk, and I. Kruci_nska, "Crystallinity of polyamide-6 matrix in glass fibre/polyamide-6 composites manufactured from hybrid yarns," *Fibres Text. East. Eur.*, 12(3), 64–69, **2004**.

28. R. L. Gorowara, W. E. Kosik, S. H. McKnight, and R. L. McCullough, "Molecular characterization of glass fiber surface coatings for thermosetting polymer matrix/glass fiber composites," *Compos. Part B*, 32, 323–329, **2001**.

29. W. J. Kim, M. Taya and M. N. Nguyen, "Electrical and thermal conductivities of a silver flake/thermosetting polymer matrix composite," *Mech. Mater.*, 41, 1116–1124, **2009**.

30. B. Koo, N. Subramanian and A. Chattopadhyay, "Molecular dynamics study of brittle fracture in epoxy-based thermoset polymer," *Compos. Part B*, 95, 433–439, **2016**.

31. S.J. Pickering, "Recycling technologies for thermoset composite materials—Current status," *Compos. Part A*, 37, 1206–1215, **2006**.

32. W. Goetz, "In: Polyamide for flexible packaging film," PLACE Conference, **2003**.

33. I. Boustead, "Eco-profiles of the European plastics industry, POLYAMIDE 66 (Nylon 66)," PlasticsEurope, **2005**.

34. S. Kuciel, P. Kuzniar and A. Liber-Knec, "Polyamides from renewable sources as matri-ces of short fiber reinforced biocomposites," *Polimery*, 57(9), 627–634, **2012**.

35. V. Selvakumar, K. Palanikumar and K. Palanivelu, "Studies on mechanical character-ization of polypropylene/Na+-MMT nanocomposites," *J. Miner. Mater. Charact. Eng.*, 9(8), 671–681, **2010**.

36. H. M. Da Costa, V. D. Ramos and M. C. G. Rocha, "Rheological properties of polypropylene during multiple extrusion," *Polym. Test.*, 24, 86–93, **2005**.

37. F. Higgins, *Determination of Percent Polyethylene in Polyethylene/Polypropylene Blends Using Cast Film FTIR Techniques*, Agilent Technologies, Danbury, CT, **2012**.

38. M. M. Kamrannejada, A. Hasanzadeh, N. Nosoudib, L. Maicand and A. A. Babaluoa, "Photocatalytic degradation of polypropylene/TiO2 nano-composites," *Mater. Res.*, 17(4), 1041, **2014**.

39. E. A. Campo, *The Complete Part Design Handbook for Injection Molding of Thermoplastics*, Carl Hanser Verlag GmbH & Co. KG, Denmark, **2006**.

40. S. Brinkmann, K. Oberbach, E. Baur, E. Schmachtenberg and T. A. Osswald, *International Plastics Handbook*, Hanser, Cincinnati, OH, **2006**.

41. V. Khudyakov, R. David Zopf and N. J. Turro, "Polyurethane nanocomposites," *Des. Monomers Polym.*, 12, 279–290, **2009**.

42. K. S. Huang, S. W. Chen, L. A. Lu and R. R. Min, "Synthesis, properties and applications of polyurethane/carboxymethyl cellulose blended polymers, I. Compatibility of the PU/CMC blended polymer," *Cellul. Chem. Technol.*, 41(2–3), 113–117, **2007**.

43. S. Ibrahim, A. Ahmad and N. S. Mohamed, "Characterization of novel castor oil-based polyurethane polymer electrolytes," *Polymer*, 7, 747–759, **2015**.

44. L. V. McAdams, J. A. Gannon and J. I. In: Kroschwitz, *High Performance Polymers and Composites*, Encyclopedia report series. John Wiley & Sons, New York, **1991**.

45. C. A. May, *Epoxy Resins Chemistry and Technology*, 2nd ed., Marcel Dekker, Inc., New York, **1988**.

46. R. Khare and S. Bose, "Carbon nanotube based composites—A review," *J. Miner. Mater. Charact. Eng.*, 4(1), 31–46, **2005**.

47. S. Bal and S. S. Samal, "Carbon nanotube reinforced polymer composites—A state of the art," *Bull. Mater. Sci.*, 30(4), 379–386, **2007**.

48. A. K. M. MoshiulAlam, M. D. H. Beg, R. M. Yunus, M. F. Mina, K. H. Maria and T. Mieno, "Evolution of functionalized multi-walled carbon nanotubes by dendritic polymer coating and their anti-scavenging behavior during curing process," *Mater. Lett.*, 167, 58–60, **2016**.

49. J. Chen, X. Cui, Y. Zhu, W. Jiang and K. Sui, "Design of superior conductive polymer composite with precisely controlling carbon nanotubes at the interface of a co-continuous polymer blend via a balance of π-π interactions and dipole-dipole interactions," *Carbon*, 114, 441–448, **2017**.

50. Z. Hu, Q. Shao, X. Xu, D. Zhang and Y. Huang, "Surface initiated grafting of polymer chains on carbon nanotubes via one-step cycloaddition of diarylcarbene," *Compos. Sci. Technol.*, 142, 294–301, **2017**.

51. S. HooshmandZaferani, "Using silane products on fabrication of polymer-based nanocomposite for thin film thermoelectric devices," *Renew. Sust. Energ. Rev.*, 71, 359–364, **2017**.

52. P. Liu, J. Han, L. Jiang, Z. Li and J. Cheng, "Polyaniline/multi-walled carbon nanotubes composite with core-shell structures as a cathode material for rechargeable lithium-polymer cells," *Appl. Surf. Sci.*, 400, 446–452, **2016**.

53. Y. Peng, K.Wang, M. Yu, A. Li and R. K. Bordia, "An optimized process for in situ formation of multi-walled carbon nanotubes in templated pores of polymer-derived silicon oxycarbide," *Ceram. Int.*, 43, 3854–3860, **2017**.

54. M. Tarfaoui, K. Lafdi and A. El Moumen, "Mechanical properties of carbon nanotubes based polymer composites," *Compos. Part B*, 103, 113–121, **2016**.

55. W. Wohlleben, "Nanorelease: Pilot interlaboratory comparison of a weathering protocol applied to resilient and labile polymers with and without embedded carbon nanotubes," *Carbon*, 113, 346–360, **2017**.

56. A. El Moumen, M. Tarfaoui and K. Lafdi, "Mechanical characterization of carbon nanotubes based polymer composites using indentation tests," *Compos. Part B*, 114, 1–7, **2017**.

57. S. Kim, J. Kissick, S. Spence and C. Boyle, "Design, analysis and performance of a polymercarbon nanotubes based economic solar collector," *Sol. Energy*, 134, 251–263, **2016**.

58. I. A. Ovid'ko, "Enhanced mechanical properties of polymer-matrix nanocomposites reinforced by graphene inclusions: A review," *Rev. Adv. Mater. Sci.*, 34, 19–25, **2013**.

59. D. Galpaya, M. Wang, M. Liu, N. Motta, E. Waclawik and C. Yan, "Recent advances in fabrication and characterization of graphene-polymer nanocomposites," *Graphene*, 1, 30–49, **2012**.

60. J. R. Potts, D. R. Dreyer, C. W. Bielawski and R. S. Ruoff, "Graphene-based polymer nanocomposites," *Polymer*, 52, 5–25, **2011**.

61. J. Li, Y. Li, S. Niu and N. Li, "Ultrasonic-assisted synthesis of phosphorus graphene oxide/poly (vinyl alcohol) polymer and surface resistivity research of phosphorus graphene oxide/ poly (vinyl alcohol) film," *Ultrason. Sono. Chem.*, 36, 277–285, **2017**.

62. R. Kausar, A. Siddiq and M. Siddiq, "Polymer/nanodiamond composites in Li-ion batteries: A review," *Polym-Plast. Technol. Eng.*, 53(6), 550–563, **2014**.

63. V. N. Mochalin and Y. Gogotsi, "Nanodiamond-polymer composites," *Diam. Relat. Mater.*, 58, 161–171, **2015**.

64. M. Chan, K. Lau, T. Wong, M. Ho and D. Hui, "Mechanism of reinforcement in a nanoclay/polymer composite," *Compos. Part B*, 42, 1708–1712, **2011**.

65. Q. T. Nguyen and D. G. Baird, "An improved technique for exfoliating and dispersing nanoclay particles into polymer matrices using supercritical carbon dioxide," *Polymer*, 48, 6923–6933, **2007**.

66. O. Zabihi, M. Ahmadi and M. Naebe, "Self-assembly of quaternized chitosan nanoparticles within nanoclay layers for enhancement of interfacial properties in toughened polymer nanocomposites," *Mater. Des.*, 119, 277–289, **2017**.

67. M. G. Sari, B. Ramezanzadeh, M. Shahbazi and A. S. Pakdel, "Influence of nanoclay particles modification by polyester-amide hyperbranched polymer on the corrosion protective performance of the epoxy nanocomposite," *Corros. Sci.*, 92, 162–172, **2015**.

68. O. Kamigaito, "What can be improved by nanometer composites?" *J. Jpn. Soc. Powder Powder Metall.*, 38, 315–321, **1991**.

69. C. M. Chan, J. Wu, J. X. Li and Y. K. Cheung, "Polypropylene/calcium carbonate nanocomposites," *Polymer*, 43, 2981–2992, **2002**.

70. Z.S. Petrovic, I. Javni, A. Waddon and G. B. Anhegyi, "Structure and properties of polyurethane–silica nanocomposites," *J. Appl. Polym. Sci.*, 76, 133–151, **2015**.

71. M. Z. Rong, M. Q. Zhang, Y. X. Zheng, H. M. Zeng, R. Walter and K. Friedrich, "Structure–property relationships of irradiation grafted nano-inorganic particle filled polypropylene composites," *Polymer*, 42, 167–183, **2001**.

72. F. Yang, Y. Ou and Z. Yu, "Polyamide 6/silica nanocomposites prepared by in situ polymerization," *J. Appl. Polym. Sci.*, 69, 355–361, **2015**.

73. Y. Ou, F. Yang and Z. Yu, "A new conception on the toughness of nylon 6/silica nanocomposite prepared via in situ polymerization," *J. Polym. Sci. Part B: Polym. Phys.* 36, 789–795, **2015**.

74. E. Reynaud, T. Jouen, C. Gauthier, G. Vigier and J. Varlet, "Nanofillers in polymeric matrix: A study on silica reinforced PA6," *Polymer*, 42, 8759–8768, **2001**.

75. C. Zeng and L. J. Lee, "Poly(methyl methacrylate) and polystyrene/clay nanocomposites prepared by in-situ polymerization," *Macromolecules*, 34, 4098–4103, **2001**.

76. Y. Li, J. Yu and Z. X. Guo, "The influence of silane treatment on nylon 6/nano-SiO2 in situ polymerization," *J. Appl. Polym. Sci.*, 84, 827–834, **2002**.

77. B. J. Ash, L. S. Schadler and R. W. Siegel, "Glass transition behavior of alumina/polymethylmethacrylate nanocomposites," *Mater. Lett.*, 55, 83–87, **2002**.

78. R. W. Siegel, S. K. Chang, B. J. Ash, J. Stone, P. M. Ajayan, R.W. Doremus and L. S. Schadler, "Mechanical behavior of polymer and ceramic matrix nanocomposites," *Scripta Mater.*, 44, 2061–2064, **2001**.

79. M. Li, Y.Q. Li and S.Y. Fu, "Cryogenic mechanical properties and thermal stability of polyimide hybrid films filled with MMT-TiO2 nano-particles," *Acta Mater. Compos. Sin.*, 23, 69–74, **2006**.

80. G. H. Hsiue, W. J. Kuo, Y. P. Huang and R. J. Jeng, "Microstructural and morphological characteristics of PS–SiO2 nanocomposites," *Polymer*, 41, 2813–2825, **2000**.

81. M. Smaihi, J. C. Schrotter, C. Lesimple, I. Prevost and C. Guizard, "Gas separation properties of hybrid imide–siloxane copolymers with various silica contents," *J. Membr. Sci.*, 161, 157–170, **1999**.

82. C. J. Cornelius and E. Marand, "Hybrid silica-polyimide composite membranes: Gas transport properties," *J. Membr. Sci.*, 202, 97–118, **2002**.

83. P. Musto, G. Ragosta, G. Scarinzi and L. Mascia, "Toughness enhancement of polyimides by in situ generation of silica particles," *Polymer*, 45, 4265–4274, **2004**.

84. Y. Zhu and D. X. Sun, "Preparation of silicon dioxide/polyurethane nanocomposites by a sol-gel process," *J. Appl. Polym. Sci.*, 92, 2013–2016, **2010**.

85. C. J. Huang, S. Y. Fu, Y.H. Zhang, B. Lauke, L. F. Li and L. Ye, "Cryogenic properties of SiO2/epoxy nanocomposites," *Cryogenics*, 45, 450–454, **2005**.

86. S. Wang, M. Wang, Y. Lei and L. Zhang, "Anchor effect" in poly(styrene maleic anhydride)/TiO2 nanocomposites," *J. Mater. Sci. Lett.*, 18, 2009–2012, **1999**.

87. Y. Kong, H. Du, J. Yang, D. Shi, Y. Wang, Y. Zhang and W. Xin, "Study on polyimide/TiO2 nanocomposite membranes for gas separation," *Desalination* 146, 49–55, **2002**.

88. Y. Li, S.Y. Fu, D.J. Lin, Y.H. Zhang and Q.Y. Pan, "Mechanical properties of polymide composites filled with SiO2 nano-particles at room and cryogenic temperature," *Acta Mater. Compos. Sin.*, 22, 11–15, **2005**.

89. M. Li, Y. Q. Li and S.Y. Fu, "Cryogenic mechanical properties and thermal stability of polyimide hybrid films filled with MMT-TiO2 nano-particles," *Acta Mater. Compos. Sin.*, 23, 69–74, **2006**.

90. D. W. Mccarthy, J. E. Mark and D. W. Schaefer, "Synthesis, structure and properties of hybrid organic–inorganic composites based on polysiloxanes. I. Poly(dimethylsiloxane) elastomers containing silica," *J. Polym. Sci. Part B: Polym. Phys.*, 36, 1167–1189, **2015**.

91. A. A. Tracton, *Coatings Technology Handbook*, 3rd ed., CRC Press, Boca Raton, FL, **2005**.

92. M. Embuscado and K. C. Huber, *Edible Films and Coatings for Food Applications*, Springer, New York, **2009**.

93. V. Rastogi, and P. Samyn, "Bio-Based coatings for paper applications," *Coatings*, 5, 887, **2015**.

94. H. Domininghaus, P. Eyerer, P. Elsner, H. T. Kunststoffe, *Eigenschaften and Anwendungen, Mit 240 Tabellen*, Springer-Verlag GmbH, Heidelberg, **2008**.

95. D. Braun, *KunststofftechnikfürEinsteiger*, Hanser, Munich, **2003**.

96. W. Kaiser, *KunststoffchemiefürIngenieure*, Hanser, Munich, **2007**.

97. H. R. Dennis, D. L. Hunter, D. Chang, S. Kim, J. L.White, J. W. Cho and D. R. Paul, "Effect of melt processing conditions on the extent of exfoliation in organoclay-based nanocomposites," *Polymer*, 42, 9513–9522, **2001**.

98. H. C. Langowski, "Permeation of gases and condensable substances through monolayer and multilayer structures." In *Plastic Packaging*, Wiley-VCH Verlag GmbH & Co. KGaA, Weinheim, 297–347, **2008**.

99. C. Bishop, *Vacuum Deposition onto Webs, Films and Foils*, Elsevier Science, Amsterdam, **2011**.

100. J. Fahlteich, "Transparente Hochbarriereschichten auf Flexiblen Substraten," Ph.D. Thesis, Technische Universität Chemnitz, Chemnitz, Germany, **2010**.

101. V. Torrisi and F. Ruffino, "Metal-Polymer nanocomposites: (co-)evaporation/(co)sputtering approaches and electrical properties," *Coatings*, 5, 378, **2015**.

102. H. Kim, A. A. Abdala, C. W. Macosko, "Graphene/polymer nanocomposites," *Macromolecules*, 43, 6515–6530, **2010**.

103. Y. S. Yoon, H. Y. Park, Y. C. Lim, K. G. Choi, K. C. Lee, G. B. Park, C. J. Lee, D. G. Moon, J. I. Han and Y. B. Kim, "Effects of parylene buffer layer on flexible substrate in organic light emitting diode," *Thin Solid Films*, 513, 258–263, **2006**.

104. E. P. Giannelis, "Polymer-layered silicate nanocomposites: Synthesis, properties and applications," *Appl. Organomet. Chem.*, 680, 675–680, **1998**.

105. P. Kiliaris and C. D. Papaspyrides, "Polymer/layered silicate (clay) nanocomposites: An overviewof flame retardancy," *Prog. Polym. Sci.*, 35, 902–958, **2010**.

106. X. Xiao-Lin, L. Qing-Xi, R. Kwok-Yiu Li, Z. Xing-Ping, Z. Qing-Xin, Y. Zhong-Zhen and M. Yiu-Wing, "Rheological and mechanical properties of PVC/CaCO3 nanocomposites prepared by in situ polymerization," *Polymer*, 45, 6665–6673, **2004**.

107. T. Adams and A. W. Charles, "Photo-oxidation of polymeric-inorganic nanocomposites: Chemical, thermal stability and fire retardancy investigations," *Polym. Degrad. Stab.*, 74, 33–37, **2001**.

108. D. S. Gary and C. L. Dimitris, "A micromechanics model for the thermal conductivity of nanotubepolymer nanocomposites," *J. Appl. Mech.*, 75(4), 1–10, **2008**.

109. P. H. C. Camargo, K.G. Satyanarayana and F. Wypych, "Nanocomposites: Synthesis, structure, properties and new application opportunities," *Mat. Res.*, 12, 1–39, **2009**.

110. L. S. Schadler, "Polymer-based and polymer-filled nanocomposites," In: *Nanocomposite Science and Technology*, Wiley-VCH, 77–153. doi:10.1002/3527602127.ch2, **2003**.

111. L. Liu and J. C. Grunlan, "Clay assisted dispersion of carbon nanotubes in conductive epoxy nanocomposites," *Adv. Funct. Mater.*, 17, 2343–2348, **2007**.

112. A. Sonia and K. Priya Dasan, "Celluloses microfibers (CMF)/poly (ethylene-co-vinyl acetate) (EVA) composites for food packaging applications: A study based on barrier and biodegradation behaviour," *J. Food Eng.*, 118, 78–89, **2013**.

113. S. S. Ray, "Recent trends and future outlooks in the field of clay-containing polymer nanocomposites," *Macromol. Chem. Phys.*, 215, 1162–1179, **2014**.

114. J. L. Suter, D. Groen and P. V. Coveney, "Chemically specific multiscale modeling of claypolymer nanocomposites reveals intercalation dynamics, tactoid self-assembly and emergent materials properties," *Adv. Mater.*, 27, 966–984, **2015**.

115. F. Alvi, M. K. Ram, P. A. Basnayaka, E. Stefanakos, Y. Goswami and A. Kumar, "Graphene-polyethylenedioxythiophene conducting polymer nanocomposite based supercapacitor," *Electrochim. Acta*, 56, 9406–9412, **2011**.

116. V. V. Radmilović, C. Carraro, P. S. Uskoković and V. R. Radmilović, "Structure and properties of polymer nanocomposite films with carbon nanotubes and grapheme," *Polym. Compos.*, 38, E490–E497, **2017**.

117. T. H. S. Maia, N. M. Larocca, C. A. G. Beatrice, A. J. De Menezes, S. G. De Freitas and L. A. Pessan, " Polyethylene cellulose nanofibrils nanocomposites," *Carbohydr. Polym.*, 173, 50–56, **2017**.

118. C. A. G. Beatrice, N. Rosa-Sibakov, M. Lille, N. Sözer, K. Poutanen and J. A. Ketoja, "Structural properties and foaming of plant cell wall polysaccharide dispersions," *Carbohyd. Polym.*, 173, 508–518, **2017**.

119. Y. Hu and J. D. Mackenzie, "Structure-Related mechanical properties of ormosils by sol-gel process," *Mater. Res. Soc. Symp. Proc.*, 271, 681, **1992**.

120. Y. J. Chung, S. Ting and J. D. Mackenzie, *Better Ceramics Through Chemistry IV*, vol. 180, Materials Research Society, Pittsburg, PA, 981, **1990**.

121. S. Anandhan and S. Bandyopadhyay, "Polymer nanocomposites: From synthesis to applications," in *Nanocomposites and Polymers* (www.intechopen.com), **2011**.

122. J. H. Koo, *Polymer Nanocomposites-processing, Characterization and Applications*, McGraw-Hill, New York, 235–261, **2006**.

123. K. Prater, "The renaissance of the solid polymer fuel cell," *J. Power Sources*, 29, 239, **1990**.

124. K. Haraguchi, "Synthesis and properties of soft nanocomposite materials with novel organic/inorganic network structures," *Polym. J.*, 43, 223–241, **2011**.

125. H. Jin, J. J. Wie and S. C. Kim, "Effect of organoclays on the properties of polyurethane/clay nanocomposite coatings," *J. Appl. Polym. Sci.* 117, 2090–2100, **2010**.

126. S.Turri, L. Torlaj, F. Piccinini and M. Levi, "Abrasion and nanoscratch in nanostructured epoxy coatings," *J. Appl. Polym. Sci.*, 118, 1720–1727, **2010**.

127. F. Ebrahimi, "Nanocomposites – New Trends and Developments," Editor InTech DTP team, Croatia, doi:10.5772/3389, **2012**.

128. G. Yu and X. Li, "Fabrication and bio-tribological properties of medical polyurethane/carbon nanotubes composites," *Energy Educ. Sci. Technol. Part A: Energy Sci. Res.*, 30(1), 475–480, **2012**.

129. C. C. Liu, S. Sadhasivam, S. Savitha and F. H. Lin, "Fabrication of multiwalled carbon nanotubes magnetite nanocomposite as an effective ultra-sensing platform for the early screening of nasopharyngeal carcinoma by luminescence immunoassay," *Talanta*, 122, 195–200, **2014**.

130. N. O. Chahine, N. M. Collette, H. Thompson and G. G. Loots, "Biocompatibility of carbon nanotubes for cartilage tissue engineering," Technical proceedings of the 2008 NSTI Nanotechnology Conference and Trade Show, NSTI-nanotech. Nanotechnology 1, 125–128, **2008**.

7 Analysis of Various Parameters on Polymer Matrix Composites

Sanam Shikalgar and Prachi Desai
Dr. Vishwanath Karad MIT World Peace University

Yashwant S. Munde
MKSSS's Cummins College of Engineering for Women

S. Radhakrishnan and Malhari Kulkarni
Dr. Vishwanath Karad MIT World Peace University

CONTENTS

7.1   INTRODUCTION

A polymer matrix composite (PMC) consists of reinforcement of fibers (short or continuous long) in a polymer resin. PMCs sustain the mechanical loads applied to structure during its service period, which is majorly due to support by reinforcements. The reinforcement offers high strength and stiffness, whereas the matrix bonds the fibers together and helps to transfer the load among them. We generally categorize PMCs into two types based on their values of strength and stiffness. First are reinforced plastics that have additional strength because of the embedded fibrous matter added into the plastic. Second are advanced composites, which consist of fiber and matrix combinations. Most of the advanced PMCs contain high-performance continuous fibers such as glass (S-glass), graphite, aramid, and carbon fibers (Allemang et al. 2014).

PMCs are light weight, corrosion resistant, and fatigue resistant compared to metals. Some general properties of PMCs enhanced along with the reinforcements are its processing, flame retardancy, fatigue resistance, fracture resistance, toughness, abrasion resistance, creep resistance, impact resistance, corrosion resistance, and thermal conductivity, which eventually reduces the overall cost. Due to these advantageous properties, it is useful in aircraft, marine, automobiles, and energy applications. Currently, PMCs restrict their usage at service temperatures below ~600°F (316°C) as the polymer decomposes at high temperatures (Agunsoye and Aigbodion 2013).

The polymer matrix, the reinforcement, and the interphase decide the properties of the PMC. Variables such as the type of matrix, type of reinforcement, constituents, physio-mechanical proportions, the geometry of reinforcement, and nature of interphase are taken into consideration while designing to produce materials that can be used in optimum conditions (Gupta and Doddamani 2018).

7.2   CONSTITUENTS OF POLYMER MATRIX COMPOSITES (PMCs)

7.2.1   Matrix

The resistance of the PMC to processes such as impact damage, layup delamination, water absorption, and high-temperature creep, which eventually cause the failure of its structure, depends on the properties of the matrix. Thermoset and thermoplastic are two types of matrix phases used in the development of PMCs (Gupta and Doddamani 2018).

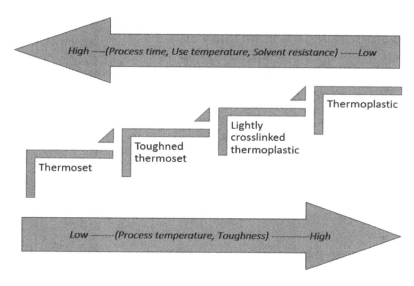

FIGURE 7.1   General characteristic of thermoset and thermoplastic matrices.

Thermosets form a three-dimensional network by connecting an entire matrix undergoing curing. Cross-linking provides high dimensional stability and high temperature resistance. It includes polymers such as polyesters, vinyl esters, epoxies, bismaleimides, and polyamides.

Thermoplastics are isolated molecules that melt to viscous liquid on heating at a temperature of 500–700°F and take the shape of the mold followed by cooling which gives rise to an amorphous, semi-crystalline, or crystalline solid. Thermoplastics include polyethylene, polycarbonate, acrylonitrile butadiene styrene (ABS), polyvinylchloride (PVC), polystyrene, etc. Reversible processing is possible in thermoplastics while thermosets are irreversible. The general characteristics of thermoset and thermoplastic matrices are shown in Figure 7.1.

7.2.2   REINFORCEMENT

In PMC, the incorporation of 60% by volume of reinforcing fibers enhances the properties of composites. Reinforcements used are glass, graphite, and aramid fibers, whereas organic fibers such as oriented polyethylene are still emerging. Natural fibers extracted from plants or animals are also promising options for the development of sustainable composites (Florea and Carcea 2012). Organic fillers affect the dimensional stability of composite rather than the polymer matrix because there is no absorption or desorption of moisture.

7.2.3   INTERPHASE

In an interphase region of PMC, transmission of loads occurs between the fiber reinforcement and polymer matrix. The level of interaction and bonding strength between matrix and reinforcement decides the interfacial ability which plays a key role in "performance" and "long-term stability" of composite against environmental and weather conditions (Florea and Carcea 2012).

7.3 ANALYSIS OF VARIOUS PARAMETERS ON POLYMER COMPOSITES

In recent years, PMCs are advantageous for various applications such as automotive, aerospace, sporting goods, marine, construction, and household. All necessary parameters need to be considered and analyzed while designing a PMC. Each of these applications are exposed to different environments, which eventually require specific properties to function effectively and efficiently. This chapter explains the effects of various parameters such as physical, chemical, mechanical, and environmental on the performance of PMC (Gupta and Doddamani 2018).

7.4 PHYSICAL PARAMETERS

Some of the physical parameters like density, particle size, aspect ratio, the volume fraction of fiber, the diameter of the fiber, nature of constituents' origin, and crystallinity of polymer influences the mechanical and thermal properties of PMCs.

7.4.1 DENSITY

The mechanical properties of a PMC improve by increasing the density of constituents of composite (fiber and matrix). Çuvalci, Erbay, and İpek (2014) investigated polyamide-based composite reinforced with short glass fibers of different densities. They showed an increase in mechanical properties like tensile strength, modulus of elasticity, and elongation at break, for 1.15 g/cm^3, 1.19 g/cm^3, 1.24 g/cm^3, and 1.3 g/cm^3 of composite density while beyond it, composites with 1.37 g/cm^3, 1.54 g/cm^3, and 1.73 g/cm^3 of densities show inversion in properties. However, it needs attention that the impact strength of composite decreases with an increase in the density of composite.

7.4.2 PARTICLE SIZE

The composite viscosity strongly depends on the particle size, distribution of particle, and surface area of the particle. Micro-sized or nano-sized fillers cause an elevation in flow activation energy of composites. Additionally, particle size may improve or reduce the mechanical properties such as Young's modulus, tensile strength, and toughness. For composites made up of large particles, reducing size leads to improvement in strength, but for smaller particles, further reduction in size does not have any effect on strength of the composite (critical particle size – usually in nano-scale).

In an experiment by Fu et al. (2008) on silica-filled PA6 nanocomposites, the authors studied the effect of 12, 25, and 50 nm particle sizes, and discovered that smaller particles (12 nm) are better reinforcement. A study conducted by Sepet, Aydemir, and Tarakcioglu (2020) on the effects of micro-sized (~2.5 μm) and nano-sized (50–100 nm) particles of $CaCO_3$ filler (10 wt.%) on HDPE reveals that ultimate tensile strength and the strain of micro-particle or nano-particle filled composite remains comparatively unchanged, similar to that of unfilled HDPE. On the other hand, the impact strength of nanoparticle-filled composite falls to 32% of impact strength of unfilled HDPE and 16% of impact strength of micro-particle-filled composite. For nanoparticle composite, they found a 5% increase in tensile modulus than unfilled HDPE and similarly 3% increase in flexural strength. On the other hand, they found an 81% increase in flexural modulus of micro-sized filler composite compared to an unfilled HDPE. They found that nano and micro-particle fillers incorporated in HDPE composite reduce its static toughness by 54% and 70%, respectively, compared to unfilled HDPE. Creep behavior of micro-particle-filled composite at low stress levels (8 and 12 MPa) is effective while at high stress levels (16 MPa), nanoparticle-filled composite improves creep behavior compared to micro-particle-filled composites. For thermal properties, nanoparticle-filled composite shows high MFI value than micro-particle-filled and unfilled HDPE composite while VST of nanoparticle and micro-particle-filled composite decreases by 2.5°C and 1°C, respectively, compared to VST of unfilled HDPE.

In another study carried out by Igwe et al. (2019) on composite made up of sugarcane bark fiber and high-density polyethylene, sugarcane bark fibers are classified into 250, 350, and 500 μm particle sizes. They found that 250 μm particle size gives maximum tensile strength, Young's modulus, and flexural strength as it absorbs more high-density polyethylene and gets properly mixed, followed by 350 and 500 μm particle sizes. Water absorption and hardness increase with an increase in particle size. This study concludes that the small particle size of fillers leads to better results.

7.4.3 ASPECT RATIO

The shape and distribution of reinforcements contribute to the properties of composites. The ratio of length to the diameter of reinforcement is the aspect ratio. Composites reinforced with continuous fibers exhibit better properties than short ones. Fibers have the highest aspect ratio compared to other forms of reinforcements such as fillers or particles (Elahi, Motevasseli, and Aghazadeh 2013). Ananthapadmanabha and Deshpande (2018) carried out a study on ABS composites made up of fillers like talc and $CaCO_3$. Talc has a high aspect ratio and orient in the flow direction, which results in a high value of properties compared to the composites made out of low aspect ratio fillers like $CaCO_3$. Thus, ABS/talc composite show better tensile and flexural strength compared to ABS/$CaCO_3$ and unfilled ABS composite while a great reduction in impact strength.

7.4.4 VOLUME FRACTION

An experiment carried out by Sathish et al. (2017) on hybrid composites reinforced with flax and bamboo fibers with different volume fractions like 0:40, 10:30, 20:20 30:10, and 40:0 (flax: bamboo) in epoxy resin. The composite containing a 40:0 volume fraction of fibers possessed high tensile and flexural strength compared to other hybrid composites formed with different amounts of volume fraction of fibers. It also gives maximum impact and interlaminar shear strength. A study by Kallyankumar,

Reddy, and Ambadas (2017) on the polyester resin glass fiber-reinforced composites showed that the composites with volume fractions of 1:99, 2:98, 3:97 (chopped glass fiber: polyester resin) raise the tensile modulus and thermal conductivity with an increase in the volume fraction of chopped glass fiber. Aramide, Oladele, and Folorunso (2009) studied bagasse fiber-reinforced polyester resin composite in which composites were prepared using 0%, 2%, 4%, 6%, 8%, 10%, 15%, and 20% of bagasse fibers as reinforcement. They showed that properties like tensile strength, modulus of elasticity, and extension to break increase with an increase in the volume fraction of fiber up to 10%, and beyond that, it begins to reduce. The strain decreases with an increase in the volume fraction of bagasse fiber in a polyester resin matrix composite.

7.4.5 CRYSTALLINITY OF POLYMER

In PMCs, the effect of reinforcement is likely to change with percentage of the crystallinity of the polymer matrix. Experimental results on the study of nanocomposites containing hydroxyapatite nanoparticles and poly-3-hydroxybutyrate matrix done by Kaur, Hoon, and Shofner (2011) found that the dispersion of nanoparticle hinders polymer crystallinity while full dispersion of nanoparticles can be achieved at low nanoparticle loading. Nanoparticle agglomeration affects reinforcement behavior in PMCs. A study of glass/polypropylene composites carried out by Wang et al. (2019) shows that the accumulation of glass fiber improves the strength of composite but reduces toughness. β-nucleating agents enhance the crystallinity along with glass fiber-reinforced polypropylene. The addition of these agents increases the toughness, impact strength, and tensile strength.

7.4.6 DIAMETER OF THE FIBER

For most materials, especially brittle materials, fibers with smaller diameters are much stronger than dense ones. Whiskers are fibers of very small size and extremely large length-to-diameter ratios and provide high crystalline perfection which is the reason for their exceptional high strength (Allemang et al. 2014). Experiments by Panigrahi, Lal Kushwaha, and Rahman (2012) on flax fiber-reinforced HDPE biocomposite focused on the study of fine fiber (19.3 μm diameter), medium fiber (26.1 μm diameter), and coarse fiber (31.6 μm diameter). They found that fine fiber at 5% of loading gives the highest tensile strength followed by medium and coarse size fibers, and at 15% of loading properties like Young's modulus, flexural strength, and flexural modulus were higher compared to medium and coarse size fiber composite.

7.4.7 NATURE OF ORIGIN

In PMCs, the matrix phase comprises thermosetting and thermoplastic resin while fibers used for reinforcement can be synthetic or natural fibers. Natural fibers are available easily, their cost is comparatively lower than that of synthetic ones, and using them as reinforcement makes them suitable for renewability and biodegradability properties. Synthetic fiber reinforcement provides high specific strength to composite, while for specific stiffness, the natural fibers are better than synthetic fibers. The most common problem associated with natural fibers in PMC is lower durability, moisture absorption, and lower impact strength (Chauhan 2017). In a study by Agunsoye and Aigbodion (2013) on bagasse-filled recycled polyethylene biocomposites, they showed that the tensile and bending strength of composites increase with the amount of bagasse fiber, and composite with 30 wt.% of bagasse fibers showed the best properties. A study of jute fiber-reinforced epoxy composite carried out by Mishra and Biswas (2013) showed that the addition of jute fibers leads to a reduction of void content in the composite, and thus 48 wt.% jute fiber content composite exhibits superior flexural and interlaminar shear strength.

7.5 CHEMICAL PARAMETERS

Chemical parameters such as the chemical composition of fiber, surface treatment of fibers, compatibilizers/coupling agents and functionalized group affect performance in terms of mechanical properties of PMCs.

7.5.1 CHEMICAL COMPOSITION OF THE FIBER

Natural fibers used as reinforcement consists of cellulose microfibrils, hemicelluloses, and lignin. On applying a load to cellulose, microfibrils align themselves with the fibers imparting an increment in its tensile strength. In a fiber, if the content of hemicellulose is high, the content of cellulose will be less. This implies an increase in moisture absorption, which tends to decrease the tensile strength of the composite. The presence of pectin imparts flexibility, and an increase in its amount leads to a decrease in tensile strength and moisture gain of the composite. The presence of lignin affects strain at the failure and increases moisture gain, whereas wax increases Young's modulus but did not affect tensile strength (Komuraiah, Shyam Kumar, and Durga Prasad 2014). They exploit the natural fibers as a replacement for conventional fiber, such as glass, aramid, and carbon due to their low cost, good mechanical properties, high specific strength, non-abrasive, eco-friendly, and biodegradability characteristics (Ku et al. 2011).

Nylon, PP, acrylic, or polyester fibers are a few of the synthetic fibers created by extruding. Usage of synthetic fibers can increase compressive strength but have a negligible influence on tensile and flexural strength. Among nylon, elastomeric, and glass fibers, nylon gives the best mechanical properties (Habib, Begum, and Mydul Alam 2013).

7.5.2 CHEMICAL TREATMENT ON THE SURFACE OF THE FIBER

The surface treatment of fibers provides a solution to various issues like hydrophilicity, thermal instability, incompatibility, and poor interfacial bonding with the matrix arising due to reinforcing polymer composites with natural fibers. It maximizes the bonding strength and eventually the stress transferability in the composites. Surface treatments of fibers are built on the usage of appropriate functional groups that alter the structure and composition of fibers resulting in the reduction of moisture absorption and eventually an increase in the compatibility with the matrix (Latif et al. 2019).

TABLE 7.1

Effect/Correlation of the Composition of Natural Fibers with Physical and Mechanical Properties

	Physical Properties		Mechanical Properties		
Chemical Composition	Density	Length and Diameter of the Fiber	Failure Strain	The Specific Strength of Fibers	Tensile Strength
Pectin	Positive	–	Positive	Positive	–
Lignin	Negative	Negative	–	Negative	Negative
Wax	Negative	Negative	–	–	–

The methods generally preferred for chemical treatments of fibers are alkaline, silane, and alkali–silane. For enhancing the properties of fiber-reinforced composites, they used coupling agents like maleic anhydride and alkali–maleic anhydride with changeable parameters such as concentration, soaking time, and the ratio by weight (Latif et al. 2019).

Alkaline (NaOH) treatment to certain natural fibers, such as textile cotton, removes some portion of fibers such as hemicellulose, lignin, pectin, and wax that makes the surface clean and void-free. This treatment leads to an increase in mechanical properties by 40%–50% as it increases the stress transfer capacity between cells. Alkali treatment of fibers removes waxy contents from the surface and increases surface roughness. It eventually raises the number of reaction sites on fiber, which results in better interlocking and improvement of mechanical properties. Alkali treatment causes cellulosic fibers to depolymerize and thus the degradation of cellulosic microfibril attributes to a decrease in flexural strength. For NaOH treatment with different concentrations, 24 hours of soaking time for alfa fiber-reinforced composite shows enhancement in its bending strength while further soaking leads to its reduction, which can be attributed to waxy substances that are removed increasing the surface roughness of fibers (Rokbi et al. 2011). Silane treatment also improves mechanical properties by increasing the fiber surface area. Along with that, increased cross-linking in it strengthens the bond between fiber and matrix. Alkali–silane treatment, on the other hand, gives 12% higher flexural strength and 7% higher modulus of composites compared to the individual alkali or silane treatment (Bodur, Bakkal, and Sonmez 2016). From a study done by Culvaci et al. for the tensile behavior of HDPE reinforced with continuous henequen fibers treated by the optimum concentration (0.015 wt.%) of silane coupling agent concentration indicated that silane increased tensile strength of the composite but none of the fiber-matrix interface improvements had any significant effect on the value of Young's modulus (Çuvalci, Erbay, and İpek 2014).

Acetylation treatment extracts waxy contents from the fiber surface and converts hemicellulose to acetyl hemicellulose partially. This results in an improvement in the mechanical properties such as tensile strength, whereas this degrades cellulosic content leading to a reduction in the tensile strength of composite and may cause internal cracks. The presence of acetyl group on flax fiber in PP composites till 18% resulted

TABLE 7.2

Effect of Different Chemical Treatments on Diverse Properties

Properties	Adverse Effects Due To	Improvement Due To
Impact strength	Alkali compared to silane; mercerization	SLS treatment compared to alkali; latex treatment, MEKP treatment
Tensile/flexural strength	Mercerization, Acetylation	–
Hydrophobicity (wettability)	–	Mercerization, acetylation, silanation, oligomeric siloxane, MAPP, coating, and graft copolymerization
Interfacial bond strength	–	Silanation, oligomeric siloxane, MAPP, coating, and graft copolymerization
The surface roughness of fibers (mechanical interlocking of fibers)	–	Mercerization, acetylation

in increasing the tensile strength while beyond it, the tensile strength decreases (Bledzki et al. 2008). Chemical treatments like acrylonitrile, silane, and methyl ethyl ketone peroxide (MEKP) on hemp fibers results in improving the impact strength, but acrylonitrile showed the best results. From the study mentioned above, it was found that the impact strength of the composite is majorly governed by the type of natural fiber rather than the type of chemical treatment used (Mehta et al. 2006).

A combination of chemical treatments like dissolution/coating/graft copolymerization can be used to treat natural fibers which improve their surface roughness, compatibility, and interfacial bond strength, leading to a substantial increase in the tensile and flexural strength of the natural plant-based fiber-reinforced composite (Latif et al. 2019). Table 7.2 summarizes all the chemical treatments performed and their effects on the properties of composites.

7.5.3 Effect of Compatibilizers/Coupling Agents on the Interfacial Adhesion between the Composite

To achieve better interfacial adhesion, materials possessing similar properties should be blended or combined (Taib and Julkapli 2018). The poor interfacial adhesion in composite again leads to inferior mechanical properties of the composite (Petinakis et al. 2009). Surface modification techniques like esterification, acetylation, and cyanoethylation are carried out on cellulosic fibers to improve the fiber-matrix interfacial interactions (Tserki, Matzinos, and Panayiotou 2006). The use of compatibilizers also improves interfacial contact, which leads to strengthening the chemical as well as physical interaction between the composites (Wang, Sun, and Seib 2002). The chemical coupling method is also one of the important chemical methods, which improve interfacial adhesion. Treating the fiber surface with a compound forms a bridge of chemical bonds between fiber and matrix. Among different coupling agents, maleic anhydride is the most commonly used as it shows improvement in tensile strength and elongation at break when grafted into matrices as compatibilizers (coupling agent). Maleated coupling agents such as MAPP (maleic anhydride PP) modify fiber surfaces and cause an increase in tensile strength and Young's modulus of about 40%, especially in PP-based composites due to enhancement in the fiber-matrix interfacial bonding, increasing the load/stress transfer capability. Composites formed without a coupling agent raises water uptake and higher diffusion coefficient compared to composites with a coupling agent. (Bodur, Bakkal, and Sonmez 2016) There is a critical amount of coupling agent for the composites, which shows the most effective interaction with fiber and matrix. The tensile properties decreased after exceeding the optimum ratio (5%) of the coupling agent.

As studied by Ku et al., fiber loading of treated (alkali and bleached) and untreated flax fiber without compatibilizer (MAPP) in PP composites cause inferior tensile strength. The treated fiber loading with compatibilizer resulted in favorable tensile strength. They found that 5 wt.% MAPP yielded the optimum value for the composites in terms of tensile strength and modulus (Ku et al. 2011). An experiment carried out by Petinakis et al. (2009) on PLA/wood flour composites shows that the addition of methylene diphenyl diisocyanate (MDI) compatibilizer results in improved tensile strength compared to the one without compatibilizer. The enhancement of the wood flour-PLA interfacial adhesion transfer stress efficiently between the constituents. A study by Ibrahim et al. (2016) showed that incorporation of 5 wt.% MAPP and nanoclay in SF/rPP (sisal fiber/recycled polypropylene) composites lead to significant improvement in tensile strength, tensile modulus, and impact strength.

7.5.4 FUNCTIONALIZATION

Functionalization causes interaction within the composite, which leads to the dispersion of molecules affecting the crystallization and thermal stability of the composite. Introducing functional groups on any part of the composite enhances bonding reaction in the matrix and builds a strong structure. In a certain experiment, where raw carbon nanotubes (CNTs) used as fillers, start agglomerating due to high surface energy and their nano size. Thus, the generation of charges on nanotube surface after acid functionalization, in which negative charges lead to a decrease in the strong self-interaction whereas the positive charges improve dispersion of CNT. The addition of functionalized nanoclay as a filler demonstrates a major role in modifying the morphology and crystallization process of the composite. The addition of Cloisite® 15A (nanoclay–MMT) enhances the thermal resistance of PHB (poly hydroxybutyrate) while the addition of Cloisite® Na^+ (functionalized) decreases it (Wang, Chen, and Tong 2006).

7.6 MECHANICAL PROPERTIES OF CONSTITUENTS

The properties of matrix and fibers along with their resultant interaction govern the properties of composites. Mechanical properties of constituents like tensile strength, elongation at break, and modulus of elasticity influence the resultant properties of the specific composite.

7.6.1 EFFECT OF THE FIBER CONTENT ON TENSILE MODULUS OF POLYMER MATRIX COMPOSITE (PMC)

The properties of composites like elastic modulus, poisons ratio, tensile strength, etc. vary linearly with the content of fibers following the rule of mixtures (Chung 2017). In general, the tensile strength of the natural fiber-reinforced polymer composites increases with fiber content, up to a maximum or optimum value. While Young's modulus for the same increase with increasing fiber loading. The tensile modulus of elasticity (E_c) of short fiber-reinforced polymer composites predicted using the modified rule of mixtures equation:

$$E_c = \lambda_E E_f V_f + E_m \left(1 - V_f\right) \tag{7.1}$$

Where λ_E is the fiber efficiency factor for the composite modulus where the effects of fiber length and orientation considered. E_f and E_m are Young's modulus of fiber and matrix, respectively.

Modulus is a property of the material at low strain and is not very sensitive to the fiber-matrix interface or critical fiber length. Hence, the fiber efficiency factor λ_E for the modulus of elasticity is relatively high than that of tensile strength (Çuvalci, Erbay, and İpek 2014). The fibers act as a load transfer medium in PMCs and helps in resisting the propagation of cracks. With an increase in fiber content, impact strength improves. Stiffness and stress transfer in composites increases with the rise in the content of fibers and thus provides a better loss and storage modulus in the dynamic properties of composites (Venkateshwaran, Elayaperumal, and Jagatheeshwaran 2011).

According to a study by Ku et al. (2011) tensile strength and Young's modulus of composites reinforced with bleached hemp fibers increased incredibly with increasing fiber loading. The mechanical properties of cast polyamide-based composites such as tensile strength and modulus of elasticity increased with increasing fiber content up to 35% fiber ratio, and after this point, increasing fiber content inversely affected these properties. Modulus of elasticity increases with the increase in tensile strength or vice versa for composite materials.

7.6.2 FIBER STRENGTH

The strength of the composites depends majorly on the strength of fibers while the latter is essentially dependent on the length of fibers. Critical fiber length (L_c) gives the effective strengthening and stiffening of the composite material. The critical fiber length is governed by parameters such as the diameter of fiber (d), an ultimate tensile strength of fiber (σ_f), and minimum fiber-matrix interfacial bond strength or shear yield strength of the matrix (T_c), as shown by Equation (7.2)

$$Lc = \frac{\sigma_f\, d}{2T_c} \tag{7.2}$$

The length of fibers significantly affects the impact strength to withstand a sudden load. The high fiber length is capable of withstanding higher bending load as well (Amuthakkannan et al. 2013). There was a test conducted by Vasiliev et al. to check the strength of a bundle of fibers (a system of approximately parallel fibers with different strengths and slightly different lengths) of two different varieties (carbon, aramid) under tension. The results showed that there is a nonlinear behavior at the initial and final stages; at initial stresses (in the vicinity of zero), nonlinearity is due to different lengths of fibers in the bundles, whereas the nonlinear behavior of the bundle under stresses close to the ultimate values is caused by fracture of the fibers with lower strength (Vasiliev and Morozov 2018).

7.7   ENVIRONMENTAL PARAMETERS

PMCs when exposed to environmental conditions like high temperature, presence of moisture, ultraviolet radiation, and corrosive chemical atmosphere start degrading. The mechanism of degradation depends upon the fiber, resin, and interphase properties of the PMC. Polymer matrix composite generally contains defects like voids, and pores caused due to processing conditions, and thus, influences the environmental degradation of PMC.

7.7.1   MECHANISM OF DEGRADATION

Figure 7.2 depicts the degradation pathways of polymeric materials. Exposure of polymeric material to the environment may cause its degradation due to biotic or abiotic factors that may act together or in sequence and leads to disintegration resulting in the formation of fragmented particles of various sizes and leached additives. Those fragmented/disintegrated particles may be of varied sizes (from macro to nano). They can interact with other potential substances and may create harmful effects on the environment.

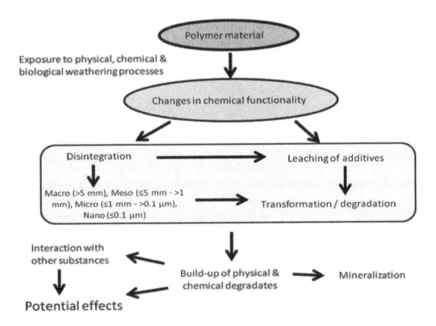

FIGURE 7.2   Degradation pathways of polymeric materials. (Scott, Alistair, and Chris 2014.)

Thus, appropriate transformation needs to be done so that they are mineralized in such a way that they can be used further or disposed off safely. The environmental factors that cause degradation and limit its usefulness by deteriorating the mechanical properties are biological attack, fatigue, temperature (UV radiation), and moisture.

7.7.2 Effect of Temperature

At elevated temperature (above 300°C), mechanical properties of composite like compressive strength, tensile strength, and stiffness decrease whereas exposure to low temperature (below 10°C) cause more brittleness and an increase in modulus of the composite. Composite exposed to cryogenic temperature becomes relatively brittle and its shear strength is bound to decrease. Thermal cycling induces micro-cracking which results in a reduction of compressive and shear strength. Experiments carried out on fiber-reinforced composites at an elevated temperature like 350°C to 450°C cause oxidation of low or high modulus fibers. A study carried out by Jayaram and Lang (n.d.) for temperature effects on fracture properties found that interlaminar fracture energy decrease with an increase in temperature.

7.7.3 Effect of Water Absorption/Moisture Penetration

In PMC, the fiber-matrix bond was stronger in the air than in the water medium. In the case of AS fiber/3002 epoxy composite studied by Muñoz and García-Manrique (2015), the composite exposed to 100% relative humidity and water immersion at 100°C shows a decrease in fracture energy of composite. Experimental findings of Han and Drzal (2003) show that in the polymer matrix based on glucose-maleic acid-ester-vinyl resin and the epoxy resin cured at high-temperature, moisture absorption causes a reduction in crosslink density leading to degradation of the polymer. High-temperature curing of polymer matrix results in brittle fracture; however, it shows better stability in a wet environment if cured at a lower temperature. In the case of basalt fiber-reinforced polymer composites, they observed a decline in tensile strength for composites immersed in wet media compared to dry specimens. They carried out immersion of specimens in two different media like normal water and saltwater, the flexural strength of specimens immersed in seawater was less compared to specimens in normal water. The increase in water absorption improves the impact strength of the composite (Amuthakkannan et al. 2013).

Absorption of water in laminate causes moisture penetration which affects fiber-matrix interface adhesion and results in debonding and generation of microcracks. This reduces the secondary interlaminar forces in composite material and it becomes more soft and ductile. Moisture absorption by resin leads to the lowering of the glass transition temperature. It also causes a reduction in mechanical properties like fracture toughness, transverse strength, and impact resistance. Moisture absorption occurs up to a saturation point and after that point, deterioration occurs (Jayaram and Lang, n.d.). An experiment carried out by Ricker, Escalante, and Stoudt (1992) on glass fiber-reinforced composite shows that when GFRPs are exposed to water at various pressure, there is a reduction in strength in all three directions. On the other hand, desorption of water causes 96% to 100% recovery.

(A)   (B)

FIGURE 7.3 Fractography of GFRP samples tested in the dry condition (a) and wet conditions (b), that is, after exposure to water for 60 days. (Ricker, Escalante, and Stoudt 1992.)

As shown in Figure 7.3, when fractography of fractures was generated initially in dry conditions and after exposure to water, initial dry fracture has areas of the relatively smooth brittle matrix while absorbed water sample area shows more ductile matrix in appearance.

7.7.4 EFFECT OF BIOLOGICAL FACTORS

The biological attack consists of fungal growth or marine fouling on a composite made up of natural fibers. The biological attack occurs because of cyclic wetting and drying and it reduces the strength of the composite (Jayaram and Lang, n.d.). The effect of biofouling on glass fiber-reinforced plastics suspended in marine water for one year at a depth of 1 m was studied by Muthukumar et al. (2011). They found that GFRP samples lost weight by 2.83% and showed 45% decrease in surface energy, 7% decrease in hardness, 19% decrease in tensile strength while elongation increased by 15%. Biological attacks on internal and external composite structures such as fiber-matrix interface are major factors contributed to the mechanical properties of composites. In studies by Brebu (2020), the fungal decay affected the dimensional stability of cellulose fiber-based polypropylene composites.

7.7.5 EFFECT OF WEATHERING

The type of climate/weathering conditions affects the performance of the composite. For fiber-glass/polyester and fiber-glass/epoxy composite, a 10%–20% decrease in tensile strength is observed which causes erosion of surface resin due to external weathering (Jayaram and Lang, n.d.). Biocomposites based on cellulose changes mechanical properties after exposure to natural weathering. Islam et al. (2013) studied biocomposites filled with cellulose from oil palm trunk and oil palm shell as nanoparticles and exposed the composite for 6 months and 12 months, respectively. They found the tensile strength to decrease from 4.8% to 11.1%, tensile modulus to decrease from 23.7% to 43%, elongation at break to decrease from 16.4% to 24.5%, flexural strength to decrease from 2.4% to 4.4%, flexural modulus to decrease from 16% to 28.3%, and impact strength to decrease from 8% to 13.3%.

7.7.6 EFFECT OF ULTRAVIOLET RADIATIONS

The degradation rate of PMC, when exposed to UV radiation depends on parameters like UV wavelength, exposure time, and UV light intensity. After UV exposure, composite material becomes fragile and its bending strength decreases by more than 50%. Generally, UV light is used for the incorporation of oxygen atoms into the polymer. The long-term UV exposure leads to a reduction in crystallinity and elasticity of composite. Kane, Mishra, and Dutta (2016) investigated the effect of UV light on jute phenolic composite and found that UV exposure for 2 years results in a decrease in tensile strength to about 50%. On exposure to UV light and humidity, degradation of cellulose, hemicellulose, and lignin content of natural fiber occurs which adversely affects the flexural strength of the composite. A study on kenaf/PET fiber-reinforced polyoxymethylene (POM) composite carried out by Nguyen et al. (2011) shows that hybrid composites retain better mechanical properties than kenaf/POM or PET/POM composite on exposure to UV radiation.

The effect of UV radiation on steel/CFRP composite was studied by Nguye et al. (2011). They found that as the temperature approaches T_g failure mode gets changed from adherend failure to debonding failure, and ultimate tensile strain of epoxy adhesive specimens decreases at a temperature near T_g. A study on thermal degradation of kenaf fiber/epoxy composite by Sugawara and Nikaido (2014) carried out at a temperature of 600°C showed that treated fiber (alkalization) lowers decomposition temperature and produces lesser char compared to untreated composite because of removal of lignin in composites.

7.8 CONCLUSIONS

Important arguments obtained from the above-described parameters such as physical, chemical, mechanical, and environmental, which eventually affect the performance of PMCs are listed below.

- Mechanical properties like tensile strength, modulus of elasticity, and elongation at break of PMC show improvement with the rise in density, aspect ratio, and volume fraction of fiber, while declining with the size of the particle.
- Suitable surface treatment of fibers and appropriate usage of coupling agents exploits the interfacial bonding, thus enhancing the strength of fiber/matrix and ultimately the stress transferability in the composites.
- Mechanical properties of constituents also play a vital role in deciding the performance of PMC. With increasing fiber content, modulus increases while tensile strength improves until an optimum level.
- Performance of composites deteriorates because of absorption of water, temperature change, or ultraviolet radiation, which are environmental conditions.

REFERENCES

Agunsoye, J.O., and V.S. Aigbodion. 2013. "Results in Physics Bagasse Filled Recycled Polyethylene Bio-Composites : Morphological and Mechanical Properties Study Q." Results in Physics 3: 187–94. doi:10.1016/j.rinp.2013.09.003.

Allemang, Randall, James De Clerck, Christopher Niezrecki, and Alfred Wicks. 2014. "Composites." Material Science and Engineering 45: 577–617.

Amuthakkannan, P., V. Manikandan, J. T. Winowlin Jappes, and M. Uthayakumar. 2013. "Effect of Fibre Length and Fibre Content on Mechanical Properties of Short Basalt Fibre Reinforced Polymer Matrix Composites." Materials Physics and Mechanics 16(2): 107–17.

Ananthapadmanabha, G. S., and Vikrant Deshpande. 2018. "Influence of Aspect Ratio of Fillers on the Properties of Acrylonitrile Butadiene Styrene Composites." Journal of Applied Polymer Science 135(11): 1–13. doi:10.1002/app.46023.

Aramide, Fatai Olufemi, Isiaka Oluwole Oladele, and Davies Oladayo Folorunso. 2009. "Evaluation of the Effect of Fiber Volume Fraction on the Mechanical Properties of a Polymer Matrix Composite." Leonardo Electronic Journal of Practices and Technologies 7(14): 134–41.

Bledzki, A.K., A.A. Mamun, M. Lucka-Gabor, and V.S. Gutowski. 2008. "The Effects of Acetylation on Properties of Flax Fibre and Its Polypropylene Composites." Express Polymer Letters 2(6): 413–22. doi:10.3144/expresspolymlett.2008.50.

Bodur, Mehmet Safa, Mustafa Bakkal, and Hasret Ece Sonmez. 2016. "The Effects of Different Chemical Treatment Methods on the Mechanical and Thermal Properties of Textile Fiber Reinforced Polymer Composites." Journal of Composite Materials 50(27): 3817–30. doi:10.1177/0021998315626256.

Brebu, Mihai. 2020. "Environmental Degradation of Plastic Composites with Natural Fillers – A Review." Polymers 12(166): 22. doi:10.3390/polym12010166.

Chauhan, Shivanku. 2017. "Study of Polymer Matrix Composite with Natural Particulate / Fiber in PMC: A Review." International Journal of Advance Research, Ideas and Innovations in Technology 3: 1168–79.

Chung, Deborah D.L. 2017. "Polymer-Matrix Composites: Mechanical Properties and Thermal Performance." In Carbon Composites, 2nd ed., 218–55. Elsevier Inc. doi:10.1016/b978-0-12-804459-9.00004-x.

Çuvalci, Hamdullah, Kadir Erbay, and Hüseyin İpek. 2014. "Investigation of the Effect of Glass Fiber Content on the Mechanical Properties of Cast Polyamide." Arabian Journal for Science and Engineering 39(12): 9049–56. doi:10.1007/s13369-014-1409-8.

Elahi, H., A.S. Motevasseli, and J. Aghazadeh. 2013. "The Influence of Aspect Ratio of Reinforcing Fibers on Mechanical Properties of Gypsum Matrix Composite Panels." CiteSeerX, no. July: 1–8.

Florea, Raluca Maria, and Ioan Carcea. 2012. "Polymer Matrix Composites - Routes and Properties." International Jornal of Modern Manufacturing Technologies IV(1): 59–64.

Fu, Shao Yun, Xi Qiao Feng, Bernd Lauke, and Yiu Wing Mai. 2008. "Effects of Particle Size, Particle/Matrix Interface Adhesion and Particle Loading on Mechanical Properties of Particulate-Polymer Composites." Composites Part B: Engineering 39(6): 933–61. doi:10.1016/j.compositesb.2008.01.002.

Gupta, Nikhil, and Mrityunjay Doddamani. 2018. "Polymer Matrix Composites." Advanced Materials by Design 70:1282–83. doi:10.1007/s11837-018-2917-x.

Habib, Ahsan, Razia Begum, and Mohammad Mydul Alam. 2013. "Mechanical Properties of Synthetic Fiber Reinforced Mortars." International Journal of Scientific & Engineering Research 4(4): 923–27.

Han, Seong Ok, and Lawrence T. Drzal. 2003. "Water Absorption Effects on Hydrophilic Polymer Matrix of Carboxyl Functionalized Glucose Resin and Epoxy Resin." European Polymer Journal 39(9): 1791–99. doi:10.1016/S0014-3057(03)00099-5.

Ibrahim, Idowu David, Tamba Jamiru, Emmanuel Rotimi Sadiku, Williams Kehinde Kupolati, and Stephen Chinenyeze Agwuncha. 2016. "Impact of Surface Modification and Nanoparticle on Sisal Fiber Reinforced Polypropylene Nanocomposites." *Journal of Nanotechnology* 2016: 9–11. doi:10.1155/2016/4235975.

Igwe, N. N., C.N. Chukwujindu, M.M. Umeji, J.C. Ibe, and N.D. Obi. 2019. "Preparation, Characterization, and Study of Effect of Particle Sizes on Different Loading of Polymer Matrix Composites Using Sugar- Cane Bark Fiber." *International Journal of Engineering Science Invention* 8(12): 71–84.

Islam, Md Nazrul, Rudi Dungani, H. P.S. Abdul Khalil, M. Siti Alwani, W. O. Wan Nadirah, and H. Mohammad Fizree. 2013. "Natural Weathering Studies of Oil Palm Trunk Lumber (OPTL) Green Polymer Composites Enhanced with Oil Palm Shell (OPS) Nanoparticles." *SpringerPlus* 2(1): 1–12. doi:10.1186/2193-1801-2-592.

Jayaram, Sreenivasa Hassan, and Jones Lang. n.d. "Impingement Of Environmental Factors That Defines A System On Composites Performance." *Engineeringcivil.Com*, 1–12.

Kallyankumar, Ramesh, Babu Reddy, and Ambadas. 2017. "Effect of Volume Fraction of Chopped Glass Fibres on Tensile and Thermal Properties of Polymer Matrix Composites." *Journal of Emerging Technologies and Innovative Research (JETIR)* 4(11): 691–96.

Kane, Shashank N., Ashutosh Mishra, and Anup K. Dutta. 2016. "UV Radiation Effect towards Mechanical Properties of Natural Fibre Reinforced Composite Material: A Review." *International Conference on Applied Sciences* 755: 9. doi:10.1088/1742-6596/755/1/011001.

Kaur, Jasmeet, Ji Hoon, and Meisha L. Shofner. 2011. "Influence of Polymer Matrix Crystallinity on Nanocomposite Morphology and Properties." *Polymer* 52(19): 4337–44. doi:10.1016/j.polymer.2011.07.020.

Komuraiah, A., N. Shyam Kumar, and B. Durga Prasad. 2014. "Chemical Composition of Natural Fibers and Its Influence on Their Mechanical Properties." *Mechanics of Composite Materials* 50(3): 359–76. doi:10.1007/s11029-014-9422-2.

Ku, H., H. Wang, N. Pattarachaiyakoop, and M. Trada. 2011. "A Review on the Tensile Properties of Natural Fiber Reinforced Polymer Composites." Composites Part B: Engineering 42(4): 856–73. doi:10.1016/j.compositesb.2011.01.010.

Latif, Rashid, Saif Wakeel, Noor Zaman Khan, Arshad Noor Siddiquee, Shyam Lal Verma, and Zahid Akhtar Khan. 2019. "Surface Treatments of Plant Fibers and Their Effects on Mechanical Properties of Fiber-Reinforced Composites: A Review." *Journal of Reinforced Plastics and Composites* 38(1): 15–30. doi:10.1177/0731684418802022.

Mehta, Geeta, Lawrence T. Drzal, Amar K. Mohanty, and Manjusri Misra. 2006. "Effect of Fiber Surface Treatment on the Properties of Biocomposites from Nonwoven Industrial Hemp Fiber Mats and Unsaturated Polyester Resin." *Journal of Applied Polymer Science* 99(3): 1055–68. doi:10.1002/app.22620.

Mishra, Vivek, and Sandhyarani Biswas. 2013. "Physical and Mechanical Properties of Bi-Directional Jute Fiber Epoxy Composites." Procedia Engineering 51: 561–66. doi:10.1016/j.proeng.2013.01.079.

Muñoz, E., and J.A. García-Manrique. 2015. "Water Absorption Behaviour and Its Effect on the Mechanical Properties of Flax Fibre Reinforced Bioepoxy Composites." *International Journal of Polymer Science* 2015: 16–18. doi:10.1155/2015/390275.

Muthukumar, Thangavelu, Adithan Aravinthan, Karunamoorthy Lakshmi, and Ramasamy Venkatesan. 2011. "International Biodeterioration & Biodegradation." International Biodeterioration & Biodegradation 65(2): 276–84. doi:10.1016/j.ibiod.2010.11.012.

Nguyen, Tien-cuong, Yu Bai, Xiao-ling Zhao, and Riadh Al-mahaidi. 2011. "Mechanical Characterization of Steel / CFRP Double Strap Joints at Elevated Temperatures." Composite Structures 93(6): 1604–12. doi:10.1016/j.compstruct.2011.01.010.

Panigrahi, Satyanarayan, Radhey Lal Kushwaha, and Anisur Rahman. 2012. "Impact of Fibre Diameter on Mechanical Properties of Flax Based Composite." American Society of Agricultural and Biological Engineers Annual International Meeting 2012, ASABE 2012 3(12): 2286–95. doi:10.13031/2013.41830.

Petinakis, Eustathios, Long Yu, Graham Edward, Katherine Dean, Hongsheng Liu, and Andrew D. Scully. 2009. "Effect of Matrix-Particle Interfacial Adhesion on the Mechanical Properties of Poly(Lactic Acid)/Wood-Flour Micro-Composites." Journal of Polymers and the Environment 17(2): 83–94. doi:10.1007/s10924-009-0124-0.

Ricker, Richard E., Emerson Escalante, and M. Stoudt. 1992. "Environmental Effects on Polymer Matrix Composites." Tri-Services Conference on Corrosion, 9.

Rokbi, Mansour, Hocine Osmani, Abdellatif Imad, and Noureddine Benseddiq. 2011. "Effect of Chemical Treatment on Flexure Properties of Natural Fiber-Reinforced Polyester Composite." Procedia Engineering, 10:2092–97 doi:10.1016/j.proeng.2011.04.346.

Sathish, S., K. Kumaresan, L. Prabhu, and N. Vigneshkumar. 2017. "Experimental Investigation on Volume Fraction of Mechanical and Physical Properties of Flax and Bamboo Fibers Reinforced Hybrid Epoxy Composites." Polymers and Polymer Composites 25(3): 229–36. doi:10.1177/096739111702500309.

Scott, Lambert, Boxall Alistair, and Sinclair Chris. 2014. "Occurrence, Degradation and Effect of Polymer-Based Materials in the Environment." Reviews of Environmental Contamination and Toxicology 227(January): vii–xi. doi:10.1007/978-3-319-01327-5.

Sepet, Harun, Bulent Aydemir, and Necmettin Tarakcioglu. 2020. "Evaluation of Mechanical and Thermal Properties and Creep Behavior of Micro- and Nano-CaCO3 Particle-Filled HDPE Nano- and Microcomposites Produced in Large Scale." Polymer Bulletin 77(7): 3677–95. doi:10.1007/s00289-019-02922-9.

Sugawara, Etsuko, and Hiroshi Nikaido. 2014. "Properties of AdeABC and AdeIJK Efflux Systems of Acinetobacter Baumannii Compared with Those of the AcrAB-TolC System of Escherichia Coli." Antimicrobial Agents and Chemotherapy 58(12): 7250–57. doi:10.1128/AAC.03728-14.

Taib, Mohamad Nurul Azman Mohammad, and Nurhidayatullaili Muhd Julkapli. 2018. "Dimensional Stability of Natural Fiber-Based and Hybrid Composites." In Mechanical and Physical Testing of Biocomposites, Fibre-Reinforced Composites and Hybrid Composites, 61–79. Elsevier Ltd. doi:10.1016/B978-0-08-102292-4.00004-7.

Tserki, V., P. Matzinos, and C. Panayiotou. 2006. "Novel Biodegradable Composites Based on Treated Lignocellulosic Waste Flour as Filler. Part II. Development of Biodegradable Composites Using Treated and Compatibilized Waste Flour." Composites Part A: Applied Science and Manufacturing 37(9): 1231–38. doi:10.1016/j.compositesa.2005.09.004.

Vasiliev, Valery V., and Evgeny V. Morozov. 2018. "Mechanics of a Unidirectional Ply." Advanced Mechanics of Composite Materials and Structures, 1–73. doi:10.1016/b978-0-08-102209-2.00001-3.

Venkateshwaran, N., A. Elayaperumal, and M. S. Jagatheeshwaran. 2011. "Effect of Fiber Length and Fiber Content on Mechanical Properties of Banana Fiber/Epoxy Composite." Journal of Reinforced Plastics and Composites 30(19): 1621–27. doi:10.1177/0731684411426810.

Wang, Hua, Xiuzhi Sun, and Paul Seib. 2002. "Mechanical Properties of Poly(Lactic Acid) and Wheat Starch Blends with Methylenediphenyl Diisocyanate." Journal of Applied Polymer Science 84(6): 1257–62. doi:10.1002/app.10457.

Wang, Shaofeng, Ling Chen, and Yuejin Tong. 2006. "Structure-Property Relationship in Chitosan-Based Biopolymer/Montmorillonite Nanocomposites." Journal of Polymer Science, Part A: Polymer Chemistry 44(1): 686–96. doi:10.1002/pola.20941.

Wang, Yuming, Lihong Cheng, Xiaoqian Cui, and Weihong Guo. 2019. "Crystallization Behavior and Properties of Glass Fiber Reinforced Polypropylene Composites." Polymers 11 (7): 14–16. doi:10.3390/polym11071198.

8 Tribo Performance Analysis on Polymer-Based Composites

Karthikeyan Subramanian
Kalasalingam Academy of Research and Education

Senthilkumar Krishnasamy
King Mongkut's University of Technology North Bangkok

Senthil Muthu Kumar Thiagamani and C. Pradeepkumar
Kalasalingam Academy of Research and Education

Aravind Dhandapani
Kalasalingam Academy of Research and
Education; Madurai Kamaraj University

Chandrasekar Muthukumar
Hindustan Institute of Technology & Science

Suchart Siengchin
King Mongkut's University of Technology North Bangkok

CONTENTS

8.1 INTRODUCTION

A few decades ago, composite materials came to the front for use in high-performing materials, structures, and many other purposes due to their specific mechanical aspects. Though the mixture of matrix and fiber increases the difficulty in design, they cause challenges in composite industries resulting in increased use of traditional materials for a given application. Though traditional materials possess many uses and advantages, a negative influence on global warming and the environment has been raised in recent days. Thus, introducing renewable and non-toxic materials such as natural fiber composites is the best way to protect the environment. Besides, natural fiber and their composites can be recycled, which is a possible way of saving resources. Therefore, many researchers and scientists have analyzed natural fiber-reinforced composites to replace conventional materials. In addition, petrochemical resources are also utilized in natural fiber-based composites [1–5]. This chapter covers the wear, erosion, and corrosion characteristics of natural fiber-reinforced polymer matrix composites.

Fiber-reinforced composites are developed based on the requirements, in particular, tribological applications. Reportedly, tribological characteristics such as wear resistance and friction coefficient are not actual assets. However, these characteristics have to function depending on the system requirements. For instance, a brake pad has to perform with a high amount of friction and less wear. In the case of bearings, it is expected to give low wear as well as low friction. The tribological characteristics are usually determined in the laboratory. However, field test is the correct choice to select the compositions for tribological applications. One of the most frequently used techniques "pin-on-disk" is employed to determine the tribological property. Equation (8.1) is used to evaluate the sliding rate of polymeric matrix composites. Besides, inversion of a specific wear rate gives the wear rate of polymeric matrix composites [6,7].

$$\text{Specific wear rate } (W_s) = \frac{\text{Loss in volume } \left(\text{cubic millimeters}\right)}{\text{Normal load} \times \text{Sliding distance } (Nm)} \tag{8.1}$$

Polymeric matrix composites are common in structural applications owing to their specific properties. The significant applications of polymeric composites are (i) rotor blades (helicopters), (ii) sand transporting pipelines, (iii) sludge transporting pipelines, (iv) plane motor fins, etc. It is observed from these applications that polymeric composites need to withstand particle erosion as all these divisions are extensively applied for surroundings with rough particles [8,9]. Many researchers have reported that the erosion behavior of polymeric matrix might be enhanced by incorporating different types of reinforcements into the matrix: nanoparticles (carbon nanotubes, nanoclays), nano microparticles (silicon carbide, boron trioxide, silicon dioxide, aluminum oxide), fibers (natural and/or synthetic), and whiskers [10–12]. Moreover, the erosion rate varies by (i) particle shape, (ii) fiber geometry, (iii) impingement angle, and (iv) impact velocity. They result in significant changes observed in the rate of material loss [13].

Understanding the behavior of fiber-reinforced composites in a harsh environment is critical. For instance, composites are subjected to chemical corrosion. Besides, the mechanism behind the chemical corrosion of composites is still unclear. Many factors are involved in fiber resistance during corrosion, such as (i) toughness,

(ii) degradation nature, (iii) resin resistance activity, and (iv) corrosion crack propagation. Fiber-reinforced composites subjected to corrosion lead to the weakening of their mechanical properties. Corrosion is exhibited in composites due to many factors, such as (i) oxidation, (ii) hydrolysis, (iii) radiation, (iv) thermal degradation, and (v) dehydration. Thus, fiber-reinforced composites must be inspected periodically to detect any corrosion and to prevent catastrophic failures. Many approaches have been used to detect the corrosion of fiber-reinforced composites: (i) nondestructive evaluation, (ii) destructive physical analysis, and (iii) mathematical technique using the finite element approach [14,15].

This chapter reviews the wear, erosion, and corrosion behavior of natural fiber-based polymeric matrix composites.

8.2 ANALYSIS OF WEAR ON NATURAL FIBER-BASED COMPOSITES

In general, polymeric matrix composites are fabricated using engineering polymers such as thermosets, thermoplastics, and elastomers. Additionally, these are used to transfer loads to the reinforcement, such as fiber or filler. Many researchers have described that the tribological characteristics of polymers are enhanced by (i) incorporating the reinforcements (with or without the fillers), (ii) varying the fiber orientations, (iii) varying the fiber architecture such as randomly or unidirectional, and (iv) fiber surface treatments. The wear analysis may be classified as sliding wear, erosive wear, adhesive wear, and abrasive wear. Figure 8.1 represents the wear articles at percentage distribution published based on the above-mentioned classifications. Figure 8.2 represents the number of articles published regarding wear analysis using natural fiber as reinforcement between 2001 and 2019.

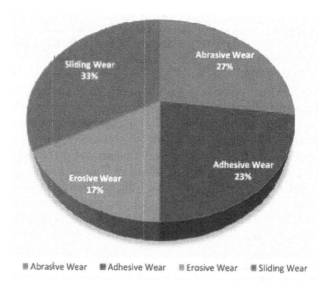

FIGURE 8.1 Wear analysis of natural fiber composites at percentage distribution [16]. (Source: http://www.sciencedirect.com; reused with permission, License number 4950580349777.)

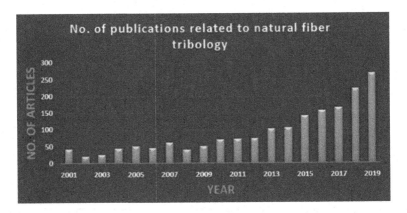

FIGURE 8.2 Number of research articles published using natural fiber as reinforcement between 2001 and 2019 [16]. (Source: http://www.sciencedirect.com; reused with permission, License number 4950580349777.)

Madnasri et al. [17] examined the wear performance of areca nut skin fiber/epoxy, pineapple leaf fiber/epoxy, and coconut fiber/epoxy composites by varying the fiber directions as: (i) random, (ii) perpendicular, and (iii) 45° orientation. Besides, the wear performance was compared by varying the fiber loadings from 2 to 10 vol%. The analysis revealed that the wear values increased on increasing the fiber loading. The coconut fiber-reinforced epoxy matrix mixtures exhibited best wear performance at 2 vol% of fiber loading between the composites. It was ascribed to the good bonding characteristics among the coconut fiber and epoxy matrix. Pradhan et al. [18] studied the abrasive wear resistance performance of newly identified fiber such as *Eulaliopsis binata*. The fiber loading was varied from 10% to 40% (order of 10%), and epoxy was used as matrix. The specific wear rate was increased in the order of: 20% < 10% < 30% < 40% < epoxy matrix, whereby the load and sliding velocity were maintained at 20 N and 0.837 m/s, respectively. When increasing the fiber loading at more than 20%, the wear performance was observed to be reduced. This reduction was ascribed to (i) the formation of fibrils and (ii) debonding between the *Eulaliopsis binata* fiber and epoxy matrix. Mercy et al. [19] analyzed the wear performance of pineapple leaf fiber-reinforced epoxy matrix composites by changing the fiber loading of 30% and 50%, respectively. Wear performance was decreased by increasing the fiber loading from 30% to 50% due to lack of fiber wetting. Nagaprasad et al. [20] incorporated *Polyalthia longigolia* seed filler in the epoxy matrix, whereby the filler sizes were varied between 25 and 50 μm. The researchers optimized the wear behavior of filler-reinforced composites by varying (i) the filler loading (0, 25, and 50 wt.%), (ii) applying load (5–15 N), and (iii) sliding speed (300–700 rpm). The polymer-based composites exhibited matrix deformation during the wear test when the filler was used in lesser quantity. On increasing the filler loading, the wear resistance was observed to high. The presence of improved filler matrix wetting ensured enhanced interfacial adhesion characteristics. Regarding the optimization of wear performance, less wear loss was observed at 25 wt.%, 10 N, and 300 rpm. The minimum coefficient of friction (COF) was observed at 25 wt.%, 5 N, and 700 rpm.

High-performance and high-temperature thermoplastic-based composites can function under (i) high temperature, (ii) high load, (iii) high speed, and (iv) severe environmental conditions. Bajpai et al. [21] compared the wear performance of three different types of natural fiber-reinforced thermoplastic-based composites such as nettle/polypropylene (PP), *Grewia optiva*/PP, and sisal/PP by varying (i) load (10–30 N), (ii) sliding speed (1–3 m/s) and (iii) sliding distance (1,000–3,000 m). Due to the abrasive nature, the sisal fiber-reinforced PP composites revealed a higher specific wear rate. Due to the differences in fiber nature and bonding characteristics in natural fibers, the specific wear rate and COF were changed at a similar sliding condition. Besides, wear performance of PP-based composites was highly influenced by applying load rather than sliding speed. Singh et al. [22] studied the COF and the specific wear rate of kenaf fiber-reinforced polyurethane composites by varying (i) fiber mat orientations, (ii) load, and (iii) sliding distances. The analysis revealed that incorporating kenaf fiber improved the wear and friction behavior of a thermoplastic matrix of about 59% and 90%, respectively. The kenaf fiber composites showed improved wear performance, and the mat was oriented 90° to the downward direction. Chand and Dwivedi [23] added maleic anhydride-grafted polypropylene (MA-g-PP) as a linking agent to the chopped jute fiber/PP composite to improve wear resistance. The linking agent helped to expand the interfacial adhesion of matrix and jute fiber during the wear process. The researchers also observed that the MA-g-PP-treated composites had higher wear resistance than the untreated and MA-g-PP melt-mixed composites. Another study exploring the COF and the specific wear rate of polylactic acid (PLA)/natural fiber-reinforced composites was done by Bajpai et al. [24]. Nettle, sisal, and *Grewia optiva* are the three types of natural fiber.

Based on the observed results, the composites did not show significant changes in the COF. A thin polymer film formation and fiber interface altered COF values. When the composites were subjected to minimum load, COF was high. It was ascribed to the mechanical interlinking of severities at the fiber interface. When the load was higher, COF was low because the thin polymer film acted as a protective layer, reducing the COF. Regarding the specific wear rate, the composites showed significant improvement than pure PLA. The wear performance of PLA-based composites was significantly influenced by applying load rather than the slipping distance. Chand et al. [25] reported that the performance of composites enhanced tribology studies, whereby the authors utilized rice husk as a filler material reinforced with polyvinylchloride (PVC). The surface-modified rice husk (using maleic anhydride) and 10 wt.% of filler-filled composites exhibited optimum tribological characteristics. When increasing the grit sizes of abrasive paper from 180 to 600 grades, wear rate was decreased. Furthermore, the wear rate of PVC matrix composites was decreased by increasing the sliding distances due to multipass abrasion. Table 8.1 lists some of the studies regarding wear of natural fiber-reinforced composites.

8.3 EROSION ANALYSIS OF NATURAL FIBER-BASED COMPOSITES

Since composite materials possess higher strength and resistance to corrosion, they are used as a replacement material for conventional materials in many applications; for instance, marine structures, construction and building industries, aerospace, and

TABLE 8.1

Research Work on Wear Analysis of Natural Fiber-Reinforced Polymer Matrix Composites

Reinforcement	Type of Matrix	Type of Wear Analysis	References
Oil palm; kenaf	Epoxy	Dry sliding	[26]
Sisal/nanosilica	Phenol-formaldehyde	Dry sliding	[27]
Palm fronds; mango dry leaves	Polyester	Dry sliding	[28]
Bamboo	Epoxy	Adhesive wear	[29]
Surface-modified palmyra fruit fiber	Polyester	Dry sliding	[30]
Sisal fiber/coconut sheath	Unsaturated polyester resin	Dry sliding	[31]
Kenaf	Epoxy	Sliding wear and friction	[32]
Betelnut	Polyester	Dry and wet sliding	[33]
Bamboo, sisal, and miscanthus	Polypropylene	Wear and friction	[34]
Jute	Polypropylene	Friction	[35]
Kenaf	Thermoplastic polyurethane	Adhesive wear	[36]
Seashell nanopowder	Polymethyl methacrylate	Dry sliding	[37]
Bagasse ash powder	Low-density polyethylene	Dry sliding	[38]
Sisal fiber, copper (I) oxide	Polylactic acid	Dry sliding	[39]

automobiles [40]. Though composite materials are used in many applications, they are subjected to damage due to erosion due to aggressive environmental conditions. Thus, many researchers have studied the erosion characteristics of polymer matrix composites based on the intended applications [41,42].

When composite surfaces are subjected to the striking of liquid or solid particles, the respective surface of the composite material undergoes removal of material. This kind of material loss can be called erosion. Many researchers have reported that the erosion behavior is controlled by several factors: (i) size of erodent, (ii) impinging velocity, (iii) impingement angle, (iv) exposure time, (v) fiber-related factors, length, loading, orientation, surface treatments, etc. [13,43]. In this section, the erosion studies on polymer matrix composites are discussed.

Getanjali et al. [44] fabricated coir/epoxy matrix composites using a hand layup technique by varying (i) impact velocity, (ii) impingement angle, (iii) fiber length, and (iv) fiber loading. The researchers further incorporated aluminum oxide fillers within the coir fiber for enhancing the wear characteristics. They reported that the wear rate improved on increasing the impact velocity irrespective of other factors. Regarding the impingement angle, the erosion rate was observed to be a peak value when the angle was 60°. Furthermore, the fiber length and fiber loading influenced the erosion behavior, whereas 12 mm fiber length and loaded composites exhibited better wear resistance behavior. Some of the important studies of erosion wear behavior of natural fiber-based composites and their significant findings are tabulated in Table 8.2.

A comparison study was reported on erosive wear studies of bamboo fiber/micro-sized red mud/epoxy and glass fiber/epoxy matrix composites [45]. The hybrid

TABLE 8.2
Tribological Analysis of Reinforced Polymer Composites

Reinforcement	Matrix	Filler/Fiber Surface Treatment	Test Conditions	Findings	References
Bamboo	Epoxy	NA	Velocity: 35–45 m/s	Peak erosions were observed at 60°–75° irrespective of fiber loading	[13]
Coir fiber	Epoxy	Aluminum oxide	Velocity: 48–109 m/s Angle: 30°–90°	Aluminum oxide filler-filled composites showed improved performance	[44]
Bamboo	Epoxy	Red mud	Velocity: 43 and 65 m/s Angle: 15°–90°	Bamboo fiber composites exhibited better performance than glass fiber composites	[45]
Arhar particulate	Epoxy	NA	Angle: 30°–90°	15% of Arhar mixed composite exhibited better erosion wear performance than the virgin epoxy matrix	[46]
Fly ash cenospheres	Epoxy	Silane treated	Angle: 30°–90°	Better performance observed at 60% of epoxy-based composites	[47]
Sugarcane bagasse fiber	Epoxy	Polyamine/ sodium hydroxide treated	Velocity: 30–70 m/s	Higher wear rate was observed with 60° impingement angles and above. Minimum wear was observed at 30 wt.% of fiber, 30 m/s, and 30°	[48]
Needle-punched nonwoven fabric	Epoxy	NA	Velocity: 43–65 m/s Angle: 30°–90°	40 wt.% of fiber-loaded composites exhibited better wear resistance, and high erosion rate was observed at 45°	[49]
Needle-punched nonwoven fabric	Epoxy	NA	Velocity: 43–65 m/s Angle: 30°–90°	Minimum wear rate was noticed at 10 wt.%, 43 m/s, and 90°	[50]

composites exhibited a significant erosive wear resistance than non-hybrid composites. Regarding the mechanical properties, the glass fiber-reinforced composites showed improved performance. However, bamboo fiber composites had higher values in terms of micro-hardness. Thus, these composites could be utilized in low load-bearing applications and can replace glass fiber-reinforced composites. Further, the Taguchi approach was used to examine erosion behavior. It was reported that the bamboo fiber composites performed better than glass fiber-reinforced composites.

In another study, the erosion rate was analyzed by varying the impingement angle (30°–90°) and fiber loading (0–40 wt.%) for bamboo/epoxy matrix composites [13]. The angle between the trajectory of the particle (before impact) and the eroded surface is called as impingement angle. Among the factors, the impingement angle could be one of the influential factors to alter the erosion behavior in polymer matrix composites. When the erosion shows a peak value with a lower impingement angle, it is concluded as "ductile mode of erosion wear." If the erosion shows a maximum value with a higher impingement angle, "brittle mode" can be assumed. Figure 8.3 shows the maximum value of erosion rate between 60° and 70° of impingement angles irrespective of fiber loadings. Thus, the behavior can be called semi-brittle and/or semi-ductile.

The influence of (i) Arhar filler loading (5–15 wt.%), (ii) impingement angles (30°–90°), and (iii) particle speeds (86, 101, and 119 m/s) on erosion wear behavior of epoxy matrix composites was studied [46]. Further, the composite samples were exposed to several environmental conditions before the erosion test: (i) dry, (ii) saline, (iii) mineral water, (iv) kerosene, and (v) subzero. The analysis revealed that the erosion behavior of Arhar-filled composites exhibited improved wear resistance than epoxy composites. Among the Arhar/epoxy composites, 15 wt.% of

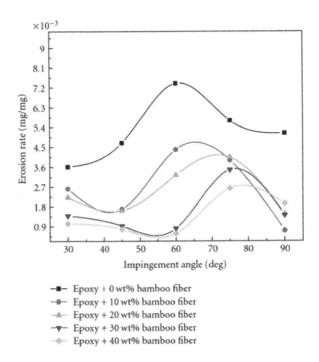

FIGURE 8.3   Erosion behavior of bamboo fiber-reinforced epoxy matrix composites [13]. (Hindawi, Open access journal.)

Arhar-filled composites did not perform well due to (i) more absorption behavior and (ii) higher erosion wear rate. However, the 10 wt.% of Arhar-filled composites could be used based on their wear behavior. Regarding the environmental conditions, the dry samples showed lesser erosion. It was ascribed to the resistance to moisture absorption behavior. Among the samples, the saline samples showed a higher erosion rate.

Shahapurkar et al. [47] formed a syntactic foam by dispersing the fly ash cenospheres (as-received and silane-treated) in the epoxy matrix. These composites were subjected to erosion wear analysis. In general, syntactic foam is preferred for lightweight structural applications. Erosion studies were conducted by varying the impact angles from 30° to 90° (order of 15°) and velocities such as 30, 45, and 60 m/s. It was reported that the erosion rate was less on increasing the impact angles and decreasing the velocity. Among the composites, the silane-treated composites showed an enhanced interfacial bonding behavior and improved erosion resistance.

It is well known that bagasse fiber is one of the major waste materials from the sugarcane industry. It is also reported that 700 million tons of bagasse fiber is generated globally [51]. Thus, many researchers have utilized bagasse fiber for their studies in polymer matrix composites. Singh et al. [48] utilized bagasse fiber for their research and studied erosion behavior by varying the control factors. The researchers used the Taguchi approach with five different control factors: (i) fiber loading, (ii) impingement angle, (iii) impact velocity, (iv) stand-off distance, and (v) erodent size to minimize the erosion wear rate. The analysis revealed that on increasing fiber content and impact velocity, the wear rate was high. Further, the bagasse/epoxy composites showed a higher erosive wear rate when the angle was 60°, which indicated a semi-brittle mode of erosion wear.

The effect of jute fiber reinforcement in the epoxy matrix and incorporation of silicon carbide (SiC) within the jute fiber/epoxy matrix composites on the erosion rate was analyzed in another study [52]. Furthermore, the authors varied the contents of jute fiber and SiC at 20–40 wt.% and 0–20 wt.%, respectively. The study showed that the erosion rate was higher at 60° for jute/epoxy matrix composites irrespective of different fiber-loaded composites. It indicated a semi-ductile mode of erosive wear. The peak impingement angle of erosion values at 0, 10, and 20 wt.% of SiC-filled composites showed 65°, 75°, and 75° respectively. This was ascribed to the transformation of ductile to brittle nature. Moreover, it is important to mention that the SiC-incorporated composites exhibited improved erosion wear resistance, whereas the weight content of the filler was higher.

A comparative study was recently reported on the erosion behavior of untreated copper slag/jute fiber/polyester and treated copper slag/jute fiber/polyester (sodium hydroxide and calcium hydroxide), having 40% of fibers with 10% of copper slag-reinforced composites [53]. The composites were subjected to erosion test by varying (i) the impingement angles (30°–90°) and (ii) jet pressure (70, 100, and 130 m/s). The analysis revealed that calcium hydroxide-treated composites exhibited a better erosion prevention than the rest of the composites. Figure 8.4 shows the erosion behavior of calcium hydroxide-treated composites, whereas the impingement angle and velocity were 60° and 100 m/s. Figure 8.4a shows the direction of scanning the surface profile. The eroded region in terms of width as well as depth is shown in Figure 8.4b.

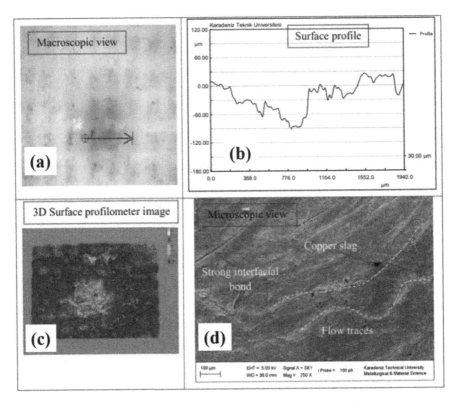

FIGURE 8.4 Erosion behavior of calcium hydroxide-treated copper slag/jute fiber/polyester composites. (Reused with permission, License number 4983041391510.)

Figure 8.4c shows that the surface behavior of calcium hydroxide-treated composites using a three-dimensional profilometer. The strong interfacial bonding between the fiber, filler, and matrix is shown in Figure 8.4d. Among the treated and untreated fiber-reinforced composites, lower erosion wear was observed for calcium hydroxide-treated composites at 90° and 86.36 m/s, respectively.

8.4   CORROSION ANALYSIS OF COMPOSITES

Nowadays, natural fiber composites are becoming a major replacement material for metal components in various industrial applications. Moreover, the performance of composite materials is improved by introducing different types of pre-treatment and hybridization techniques. Due to its moisture absorbing capacity, natural fibers does not exhibit strength in particular situations. Thus, pre-treatment and hybridization play a major role in composite manufacturing.

The corrosion resistance properties of glass, carbon, and basalt fibers with natural fibers provide added advantages for the composite. In this manner, the corrosion resistance properties of the composites could be improved. Of the various natural and synthetic fibers, carbon and basalt fibers are used as major corrosion resistance materials [54]. Corrosion of composite is profound in many situations, particularly in marine and oilfield environments, which might be considered for high temperature, corrosive situation, and high strain [55]. A corrosion-resistant material should have the following properties: (i) better durability and adhesion, (ii) good surface finish, (iii) maximum resistance properties, (iv) and cost-effectiveness. Figure 8.5 shows the major classification of anticorrosion coating.

Some of the issues related to corrosion of the materials are elaborated below.

Uniform dispersion: In numerous instances, filler or dispersing fabric familiar with reinforcing steel or polymer matrix provides predicted reinforcement because of the collection of fillers. It ensures greater coating properties.

Novel coating material: Coating materials with excessive mechanical stability and corrosion resistance need to be investigated. A material that has excessive corrosion resistance may have less mechanical stability and vice versa. Increasing growth of hybrid nanocomposite coatings with promising applications can solve the corrosion problem without losing the surface mechanical homes of the coating.

Advanced tools: Relationships between the microstructure of coating materials can be fully acknowledged most effectively through progressive characterization techniques, which include high temperature, nanoindentation, atom probe tomography, and synchrotron X-ray nanodiffraction [56].

Vijayan et al. [57] studied the protection of carbon steel from corrosion. For this, chicken feather fiber (CFF) epoxy coating was developed. Erosion concentrates on the coatings were assessed utilizing a quickened salt drenching test. CFF was dispersed on the coating surface uniformly. The authors found that CFF reduced the wettability of the coating surface and confirmed that CFF could act as good corrosion resisting agent for epoxy coating on a carbon steel substrate. In another study, they prepared short basalt fiber-reinforced aluminum matrix composite by vacuum hot-press sintering process. Basalt short fibers were uniformly distributed on the aluminum matrix. After the addition of short basalt fibers, the transfer film formed during the corrosion process of the composites was stronger and denser due to the formation of the Al_2O_3 reaction zone, thus improving corrosion resistance. Furthermore, the corrosion behavior of composites was investigated by hydrogen evolution and electrochemical tests. The results showed that the corrosion resistance of 7075 aluminum

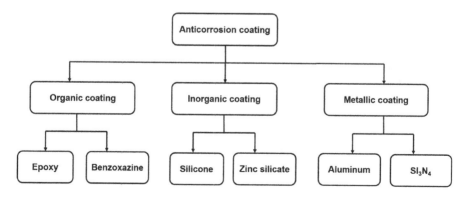

FIGURE 8.5 Classification of anticorrosion coating.

alloy was significantly improved after adding 1.0 wt% of short basalt fiber [58]. Jiang et al. [59] studied the corrosion wear resistance of wood plastic composite under simulated environmental conditions. The simulated conditions included seawater and acid rain environment. The most damaging corrosion condition was evaluated using the simulation. They show that seawater at 55°C with a salt content of 3.5% and acid rain with a PH value of 2.5 could damage the sample at a higher rate. The prolonged usage of this composite under these conditions will damage the composite by reducing its mechanical and thermal properties. A similar test was also conducted with the addition of microsilica (MS) and acrylonitrile styrene acrylate (ASA) as filler materials. The corrosion resistance of the material increased on the addition of filler material. The addition of 6% of MS and 34% of ASA showed better corrosion and wear resistance property of the composite. It also contributed to the mechanical property [60]. Sendil et al. [61] fabricated a hand layup composite using glass, jute, and nylon fiber along with epoxy as a matrix for boat structural application. The corrosion resistance of these composites was tested because it was used in the seawater environment. Tests such as saltwater spray and corrosive liquid immersion were performed in the composite. The results revealed that this composite could withstand the corrosive environment for a prolonged period. Jute and nylon absorbed some water during immersion test and increased the weight of the composite by 1%. Under seawater testing conditions, many researchers have tested alternative materials for marine application. Similarly, polymers were tested for their corrosion behavior under saltwater and plain water conditions. The corrosion behavior of the polyether ether ketone (PEEK), polyamide (PI), perfluroethylene propylene copolymer (FEP), and some other polymer materials was tested in these environments. All materials were sliding against the steel with seawater lubrication. The results revealed that PEEK, PI, and FEP showed better COF and wear rate at seawater conditions [62]. Very few researchers have worked on the improvement of anticorrosion behavior of natural fiber composite material. The scope of expanding the research on this area is clear, and different coating mechanisms also need to be explored.

8.5   CONCLUSIONS

In this chapter, recent developments related to natural fiber-based composites and their importance in tribological, erosive, and corrosion-related works were discussed. The wear characteristics of natural fiber-based composites were highlighted based on different operating conditions, such as sliding distance, sliding velocity, and load application. The inclusion of reinforcements in the form of fibers or fillers or nanoparticles to the polymer matrix can modify the material's surface characteristics, resulting in improved tribological and erosion behaviors. A better erosion characteristic could be observed by having a normal fiber direction against the sliding direction. Corrosion resistance studies were also reviewed in this chapter. It was revealed that the coating of natural fiber and hybridization are better options to improve this property. Addition of synthetic polymer material along with natural material has also increased recently.

However, the research is not exhaustive. There are many unexplored natural fibers, plants, and fillers that need to be studied. It would help bring a remarkable development in the field of tribology and address the issues in several fields.

ACKNOWLEDGEMENTS

The authors wish to thank Kalasalingam Academy of Research and Education, Krishnankoil 626126, India. This research was also supported by King Mongkut's University of Technology North Bangkok (KMUTNB), Thailand through Grant No. KMUTNB-64-KNOW-07.

REFERENCES

1. Thiagamani SMK, Krishnasamy S, Siengchin S (2019) Challenges of biodegradable polymers: An environmental perspective. *Appl Sci Eng Prog* x:1. https://doi.org/10.14416/j.asep.2019.03.002
2. Campilho RDSG (2015) *Natural Fiber Composites*. CRC Press, Boca Raton, FL.
3. Pickering K (2008) *Properties and Performance of Natural-fibre Composites*. Elsevier, Amsterdam.
4. Senthilkumar K, Kumar TSM, Chandrasekar M, et al. (2019) Recent advances in thermal properties of hybrid cellulosic fiber reinforced polymer composites. *Int J Biol Macromol* 141:1–13.
5. Senthilkumar K, Siva I, Rajini N, Jeyaraj P (2015) Effect of fibre length and weight percentage on mechanical properties of short sisal/polyester composite. *Int J Comput Aided Eng Technol* 7:60–71.
6. Friedrich K, Zhang Z, Schlarb AK (2005) Effects of various fillers on the sliding wear of polymer composites. *Compos Sci Technol* 65:2329–2343. https://doi.org/10.1016/j.compscitech.2005.05.028
7. Karthikeyan S, Rajini N, Jawaid M, et al (2017) A review on tribological properties of natural fiber based sustainable hybrid composite. *Proc Inst Mech Eng Part J J Eng Tribol* 231:1616–1634.
8. Bagci M, Demirci M, Sukur EF, Kaybal HB (2020) The effect of nanoclay particles on the incubation period in solid particle erosion of glass fibre/epoxy nanocomposites. *Wear* 444:203159.
9. Papadopoulos A, Gkikas G, Paipetis AS, Barkoula N-M (2016) Effect of CNTs addition on the erosive wear response of epoxy resin and carbon fibre composites. *Compos Part A Appl Sci Manuf* 84:299–307.
10. Sun X, Wang Y, Li DY, et al (2014) Solid particle erosion behavior of carbidic austempered ductile iron modified by nanoscale ceria particles. *Mater Des* 62:367–374.
11. Chen J, Trevarthen JA, Deng T, et al (2014) Aligned carbon nanotube reinforced high performance polymer composites with low erosive wear. *Compos Part A Appl Sci Manuf* 67:86–95.
12. Mishra SK, Biswas S, Satapathy A (2014) A study on processing, characterization and erosion wear behavior of silicon carbide particle filled ZA-27 metal matrix composites. *Mater Des* 55:958–965.
13. Gupta A, Kumar A, Patnaik A, Biswas S (2011) Effect of different parameters on mechanical and erosion wear behavior of bamboo fiber reinforced epoxy composites. *Int J Polym Sci* 2011:1–11.
14. Abidin NIZ, Sabri MFM, Kalantari K, et al. (2019) Corrosion detection for natural/synthetic/textiles fiber polymer composites. In: *Structural Health Monitoring of Biocomposites, Fibre-reinforced Composites and Hybrid Composites*. Elsevier, pp. 93–112.
15. Shan Y, Liao K (2001) Environmental fatigue of unidirectional glass–carbon fiber reinforced hybrid composite. *Compos Part B Eng* 32:355–363.
16. Chaudhary V, Ahmad F (2020) A review on plant fiber reinforced thermoset polymers for structural and frictional composites. *Polym Test* 91:106792.

17. Madnasri S, Astika G, Marwoto P (2020) The effects of natural fiber orientations on the mechanical properties of brake composites. *J Nat Fibers* 17:1–12.

18. Pradhan S, Acharya SK, Prakash V (2020) Mechanical, morphological, and tribological behavior of Eulaliopsis binata fiber epoxy composites. *J Appl Polym Sci* 50077.

19. Mercy JL, Parmar DS, Srivastava S (2020) Tribological and thermogravimetric analysis of pineapple fibre reinforced epoxy composite. In: *IOP Conference Series: Materials Science and Engineering.* IOP Publishing, p. 12024.

20. Nagaprasad N, Stalin B, Vignesh V, et al. (2020) Applicability of cellulosic-based Polyalthia longigolia seed filler reinforced vinyl ester biocomposites on tribological performance. *Polym Compos* 42:791–804.

21. Bajpai PK, Singh I, Madaan J (2013) Frictional and adhesive wear performance of natural fibre reinforced polypropylene composites. *Proc Inst Mech Eng Part J J Eng Tribol* 227:385–392. https://doi.org/10.1177/1350650112461868

22. Singh N, Yousif BF, Rilling D (2011) Tribological characteristics of sustainable fiber-reinforced thermoplastic composites under wet adhesive wear. *Tribol Trans* 54:736–748.

23. Chand N, Dwivedi UK (2006) Effect of coupling agent on abrasive wear behaviour of chopped jute fibre-reinforced polypropylene composites. *Wear* 261:1057–1063.

24. Bajpai PK, Singh I, Madaan J (2013) Tribological behavior of natural fiber reinforced PLA composites. *Wear* 297:829–840. https://doi.org/10.1016/j.wear.2012.10.019

25. Chand N, Sharma P, Fahim M (2010) Tribology of maleic anhydride modified rice-husk filled polyvinylchloride. *Wear* 269:847–853.

26. Shuhimi FF, Abdollah MF, Bin, Kalam MA, et al. (2016) Tribological characteristics comparison for oil palm fibre/epoxy and kenaf fibre/epoxy composites under dry sliding conditions. *Tribol Int* 101:247–254.

27. Wei C, Zeng M, Xiong X, et al. (2015) Friction properties of sisal fiber/nano-silica reinforced phenol formaldehyde composites. *Polym Compos* 36:433–438.

28. Ibrahim RA (2015) Tribological performance of polyester composites reinforced by agricultural wastes. *Tribol Int* 90:463–466.

29. Nirmal U, Hashim J, Low KO (2012) Adhesive wear and frictional performance of bamboo fibres reinforced epoxy composite. *Tribol Int* 47:122–133.

30. Irullappasamy S, Durairaj R, Irulappasamy S, Manoharan T (2018) Investigation on wear behaviors and worn surface morphology of surface treated palmyra fruit fiber/polyester composites to appraise the effects of fiber surface treatments. *Polym Compos* 39:2029–2035.

31. Chandrasekar M, Senthilkumar K, et al. (2021) Effect of adding sisal fiber on the sliding wear behavior of the coconut sheath fiber-reinforced composite. *Tribol Polym Compos Charact Prop Appl.* 115–125.

32. Chin CW, Yousif BF (2009) Potential of kenaf fibres as reinforcement for tribological applications. *Wear* 267:1550–1557.

33. Nirmal U, Yousif BF, Rilling D, Brevern PV (2010) Effect of betelnut fibres treatment and contact conditions on adhesive wear and frictional performance of polyester composites. Wear 268:1354–1370. https://doi.org/10.1016/j.wear.2010.02.004

34. Chegdani F, Mezghani S, El Mansori M, Mkaddem A (2015) Fiber type effect on tribological behavior when cutting natural fiber reinforced plastics. *Wear* 332:772–779.

35. Yallew TB, Kumar P, Singh I (2014) Sliding wear properties of jute fabric reinforced polypropylene composites. *Procedia Eng* 97:402–411.

36. Narish S, Yousif BF, Rilling D (2011) Adhesive wear of thermoplastic composite based on kenaf fibres. *Proc Inst Mech Eng Part J J Eng Tribol* 225:101–109.

37. Karthick R, Sirisha P, Sankar MR (2014) Mechanical and tribological properties of PMMA-sea shell based biocomposite for dental application. *Procedia Mater Sci* 6:1989–2000.

38. Aigbodion VS, Hassan SB, Agunsoye JO (2012) Effect of bagasse ash reinforcement on dry sliding wear behaviour of polymer matrix composites. *Mater Des* 33:322–327.
39. Senthilkumar K, Siengchin S, Senthil Muthu Kumar T, et al. (2020) Tribological characterization of cellulose fiber-reinforced polymer composites. *Tribol Polym Compos Charact Prop Appl* 95–113.
40. Chandrasekar M, Siva I, Kumar TSM, et al. (2020) Influence of fibre inter-ply orientation on the mechanical and free vibration properties of banana fibre reinforced polyester composite laminates. *J Polym Environ* 28:2789–2800.
41. Patnaik A, Satapathy A, Chand N, et al. (2010) Solid particle erosion wear characteristics of fiber and particulate filled polymer composites: A review. *Wear.* https://doi. org/10.1016/j.wear.2009.07.021
42. Abdul-Hussein AB, Hashim FA, Kadhim TR (2015) Comparison study of erosion wear and hardness of GF/EP with nano and micro SiO2 hydride composites. *Eng Technol J* 33:1761–1774.
43. Suresha B, Siddaramaiah, Kishore, et al. (2009) Investigations on the influence of graphite filler on dry sliding wear and abrasive wear behaviour of carbon fabric reinforced epoxy composites. *Wear* 267:1405–1414. https://doi.org/10.1016/j.wear.2009.01.026
44. Das G, Biswas S (2017) Erosion wear behavior of coir fiber-reinforced epoxy composites filled with Al 2 O 3 filler. https://doi.org/10.1177/1528083716652832
45. Biswas S, Satapathy A (2010) A comparative study on erosion characteristics of red mud filled bamboo – epoxy and glass – epoxy composites. *Mater Des* 31:1752–1767. https://doi.org/10.1016/j.matdes.2009.11.021
46. Prakash MO, Raghavendra G, Panchal M, et al. (2017) Effects of environmental exposure on tribological properties of arhar particulate / epoxy composites. *Polym Compos* 39:3102–3109. https://doi.org/10.1002/pc
47. Shahapurkar K, Doddamani M, Kumar GCM, Gupta N (2019) Effect of cenosphere filler surface treatment on the erosion behavior of epoxy matrix syntactic foams. *Polym Compos* 40:2109–2118. https://doi.org/10.1002/pc.24994
48. Singh T, Tejyan S, Patnaik A, et al (2019) Fabrication of waste bagasse fiber-reinforced epoxy composites: Study of physical, mechanical, and erosion properties. *Polym Compos* 40:3777–3786. https://doi.org/10.1002/pc.25239
49. Tejyan S, Patnaik A (2015) Erosive wear behavior and dynamic mechanical analysis of textile material reinforced polymer composites. *Polym Compos* 38:2201–2211. https:// doi.org/10.1002/pc
50. Tejyan S, Singh T, Patnaik A, Fekete G, Gangil B (2018) Physico-mechanical and erosive wear analysis of polyester fibre-based nonwoven fabric-reinforced polymer composites. *Journal of Industrial Textiles* 49. https://doi. org/10.1177/1528083718787524
51. Ahmed AESI, Hassan ML, El-Masry AM, El-Gendy AM (2014) Testing of medical tablets produced with microcrystalline cellulose prepared from agricultural wastes. *Polym Compos.* https://doi.org/10.1002/pc.22786
52. Mantry S, Satapathy A, Jha AK, et al. (2011) Preparation, characterization and erosion response of jute-epoxy composites reinforced with SiC derived from rice husk. *Int J Plast Technol* 15:69–76. https://doi.org/10.1007/s12588-011-9007-z
53. Kalusuraman G, Thirumalai Kumaran S, Aslan M, et al. (2019) Use of waste copper slag filled jute fiber reinforced composites for effective erosion prevention. Meas J Int Meas Confed. https://doi.org/10.1016/j.measurement.2019.106950
54. Amalina MA, Ahmad R (2019) Corrosion detection for natural/synthetic/textiles fiber polymer composites. In *Structural Health Monitoring of Biocomposites, Fibre-Reinforced Composites and Hybrid Composites.* https://doi.org/10.1016/ B978-0-08-102291-7.00006-X

55. Raja S, Ravichandran M, Stalin B, Anandakrishnan V (2020) A review on tribological, mechanical, corrosion and wear characteristics of stir cast AA6061 composites. *Mater Today Proc* 22:2614–2621. https://doi.org/https://doi.org/10.1016/j.matpr.2020.03.392

56. Sahoo P, Das SK (2011) Tribology of electroless nickel coatings – A review. *Mater Des* 32:1760–1775. https://doi.org/https://doi.org/10.1016/j.matdes.2010.11.013

57. Vijayan PP, Bhanu AVA, Archana SR, et al. (2020) Development of chicken feather fiber filled epoxy protective coating for metals. *Mater Today Proc.* https://doi.org/https://doi.org/10.1016/j.matpr.2020.05.229

58. Ma Y, Xiong H, Chen B (2020) Preparation and corrosion resistance of short basalt fiber/7075 aluminum composite. *Mater Corros* 71:1824–1831. https://doi.org/https://doi.org/10.1002/maco.202011771

59. Jiang L, He C, Fu J, Chen D (2017) Wear behavior of wood–plastic composites in alternate simulated sea water and acid rain corrosion conditions. *Polym Test* 63:236–243.

60. Jiang L, He C, Fu J, Xu D (2019) Enhancement of wear and corrosion resistance of polyvinyl chloride/sorghum straw-based composites in cyclic sea water and acid rain conditions. *Constr Build Mater* 223:133–141. https://doi.org/10.1016/j.conbuildmat.2019.06.216

61. Sendil Kumara R, Dinesha P, Vigneshb G, Vikin Rajb B, Vinoth Rajb V, Ramkumar S (2018) Experimental investigation of corrosive wear of boat directional. *Int J Sci Res Innov* 5:1–6.

62. Chen B, Wang J, Yan F (2011) Friction and wear behaviors of several polymers sliding against GCr15 and 316 steel under the lubrication of sea water. *Tribol. Lett* 42:17–25. https://doi.org/10.1007/s11249-010-9743-9

9 Failure Analysis of Polymer-Based Composites

N. Sabarirajan
Chendhuran College of Engineering and Technology

T. Sathish and R. Deepak Joel Johnson
Saveetha School of Engineering, SIMATS

CONTENTS

9.1   INTRODUCTION

Composites have two categories, namely, advance and engineering composites. In addition to these two categories, there are several polymer composites, which are made up of thermoplastic resin or thermoset matrices. Polymer composites combine two material, in which the base material is a polymeric resin. Polymer composites have similar characteristics as metal matrix composite and ceramic composites. These composites are used to alter the properties of the base material. Nowadays, polymer composites are ubiquitous, and are widely used in aerospace, marine, electronics, wind energy, sports goods, etc.

Polymer composite consists of two major phases, namely, continuous and discontinuous phase. The material dominating in continuous phase is resin, which is also termed as matrix or polymer phase. On the other hand, the discontinuous phase is dominated by the reinforcement material such as glass, silicon fillers, carbon fibres, etc.

The polymer-based composition is lightweight, has high transition temperature, tailor properties, good fatigue resistance, and so on. The polymer composite also has a few disadvantages, such as high cost of raw materials, complex manufacturing process, and low thermal expansion. However, the polymer composite results in synergistic mechanical properties, which are not easily attained by a single component.

Polymer composites are widely used in high-performance engineering applications. However, the earlier researchers were unaware about the defect sensitivity. The defect can occur due to the poor quality and design of material. Moreover, in earlier stages, industries did not understand how to handle the composite for their product. Due to these knowledge gaps, failure was a common problem. Thus, it is essential to analyse the failure of polymer composite before usage.

Failure in polymer composite can classified into the following types: material defect, design failure, in-services anomalies, and manufacturing defects. Hence, computational analysis is recommended to analyse the materials and composite. One of the noticeable analysis is dynamic and static loadings. Loading can help to find determine failure of composite, and practical criteria are employed to predict the failure. Based on the criteria applied, the accuracy of failure prediction can also vary.

9.2   FAILURE PREDICTION

Failure prediction is an essential computational analysis used to forecast the life of the composite component. The traditional failure prediction criteria vary form the simplest noninteractive to the interactive (Berthelot 1999). Recently, modern failure prediction criteria have emerged, which can analyse the matrix and fibre failures. The fibre failure can be tested based on tensile and compressive strength tests. On the other hand, the matrix failure can be analysed by direct mode criteria. In some special types of failure, delamination occurs, in which layers get separated. Delamination occurs due to shear stress or normal stress between layers. This type of failure is caused either by normal stress components between the layers or by shear stresses acting in a plane of layers (Laš et al. 2006).

Tensile Strength: Tensile strength defines the weight required to break a seam, divided by the cross-sectional area of the suture. The relationship between the weight and beam diameter required to break a beam is not linear (ASM Handbook 1987). Tensile strength can be measured using dry or wet formulas. The tensile strength of wet formulas is clinically most applicable when the formulas are applied to tissues with extracellular matrix. Effective tensile strength is an additional measure that measures the tensile strength of a bent and knotted fabric. The effective tensile strength varies with the material and the type of knot. A nucleus must have sufficient tensile strength for specific purposes.

Compressive Strength: Compressive strength or compression strength is the capacity of a material or structure to withstand loads tending to reduce size, as opposed to which withstands loads tending to elongate. In other words, compressive strength resists being pushed together, whereas tensile strength resists tension.

Puck's Failure Criterion: The Puck fracture criteria for unidirectional fibre/polymer composites are widely accepted for physical prediction of failure and post-failure degradation behaviour. The core of the theory, the inter-fibre fracture criteria, are criteria of the Coulomb/Mohr type. Accordingly, the inclination of the fibre-parallel fracture plane – in other words, the fracture angle – has to be determined in the first analysis step. Afterwards, the fracture stresses acting on the predicted fracture plane can be determined. Since the theory was first introduced, continuous efforts have been taken to reduce the computational cost of such two-step approaches.

9.3 METHODS TO ANALYSIS FRACTURE MECHANISM

The hazardous materials are a set of disaster components and that can lead to the analysis the system. Therefore, proctographic analysis is important to understanding the risks associated with the composite materials and to mitigate them. The risks are then avoided and are briefly explained here. In the complex repair manuals, the available information gets more complex (Dorworth 2001; Greenhalgh and Hiley 2008; Kar 1992; Armstrong et al. 2005). Using the handling of composite material, the health issues can be divided into three set levels based on severity. The level of the security such as an isolated environment, and in-service failures. These failures are showing to the fire of the modules (Greenhalgh 2009). Also, the risks are separated and associated with the behaviour of the composite components. The risk associated with the components is related to the cutting and machining of the system.

The study of failed components is analysed and provided for the materials. Then, the specific hazard connects to the incident. Also, the search of the materials is easily undertaken to allow only personnel and to handle the materials for data protection. These materials also decrease the safety as well as the health risks through the failure (Greenhalgh 2009). Some points related to running of the failed mixtures include not eating or drinking in the locality of the components. Finally, the analysis of failed component can be sensible and which can confirm the tetanus of inoculated skin.

* Handing of Single Component from an Isolated Environment

The failure of hazardous components is the mechanical step for defective production material in controlled or isolated environment. The important elements of components include loose fibres, dust, and splinters. The person should wear rubber gloves to prevent skin punctures. The splinters of glass as well as carbon fibre tend to break and are removed. During all stages, the specimen needs to be handled properly to avoid injury.

* Handing of Failed In-Service Components

In the next stage, the accident is connected to the service of failures, for example, a failed component of vehicles or faulty aircraft. Loose fibres and crackers are the most dangerous which can also pollute the surrounding system. Under these conditions, the fuel or hydraulic system is suggested to appropriate personnel along with the safety equipment.

During an accident, based on the injuries the failure service is coming to the biological hazards. Guidelines should be followed according to the standard laboratory practice.

9.3.1 PROCEDURE FOR ANALYSIS

The failure analysis resolves the associated large amount of information and requires minimal cost resources (McCoy 2004). During the analysis, contradictory claims lead to difficulties.

With scientific rigour, the procedure confirms the study of sound and reinforced. The procedure should maximise the amount of information and fractographic. The quantity of data is misplaced and confirm the investigation was as efficient as possible. In laminated polymers, the required time is the behaviour of failure analysis. They are frequently larger than the metallurgical samples, and the fracture surface area generates the interlinear composites.

9.3.1.1   Overall Procedure

The characteristic procedure is shown in Figure 9.1. The virtual importance of the detailed steps is applicable for exploring the polymer composite failures (Becker et al., 2002; Dorworth 2001; Grove and Smith 1987; Greenhalgh and Hiley 2008; Lemascon et al., 1996). The procedure is significant for analysis, as shown in Figure 9.1. In the separation process, it is important to collect the information and disconnect to the backlink sources.

Fractographic analysis is important for the end users, and it cannot analyse the surface of the components. The information is contributing to the gleaned sequences and sources like as a material or processing quality, failure, evidence of corrosion

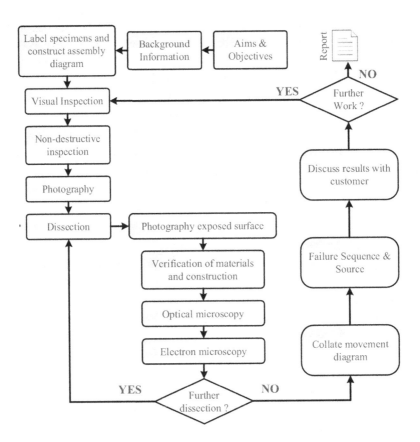

FIGURE 9.1   Overall procedure of failure analysis.

or wear and perhaps factors. However, the quantifiable information is not directly determined fractographically. The user should be aware of the preliminary outcomes which are subject to the alteration for the subsequent analysis. Finally, the work should be decided for the format for reporting and the timescales of investigation.

The users are request to refer the failure and it can change in the future. The range of the failure analyst should only adequate component of the stress and design of the composite materials. The next stage must include the collection of background information and the production of failed components (Dorworth 2001; Greenhalgh and Hiley 2008). These components include data such as a fibre stacking sequence, resin combinations, and detailed geometries. In particular, they are different from the manufacturer-suggested procedures. The service history includes data such as loading conditions, service environment, and repairs. Figure 9.2 provides a detailed description of generation of the relative data.

The fracture surface is observed using an optical and electron microscope. The figure should validate the dissimilar fracture patterns as well as the guidelines of failures on the crack path (Quinn and Quinn 2010). The operating maps also classify sites such as the boundary among the modes of failures visualised using an electron microscope. A compound microscope provides additional information on the process of fracture.

In the fracture surface, the motion diagram can be drawn from digital photography (Bonhomme et al. 2009; Purslow 1984). It is useful to place a scale on the motion chart to allow comparisons between features in different motion charts. The typical labelling for referring to such motion charts is shown in Table 9.1.

9.3.1.2 Post-failure Damage

Figure 9.1 shows the steps involved in post-failure damage. In the handling fracture, the joints of failure are followed by the procedure of safety and hygiene. Most damages and artefacts present on the surface of the fracture. Damages are classified

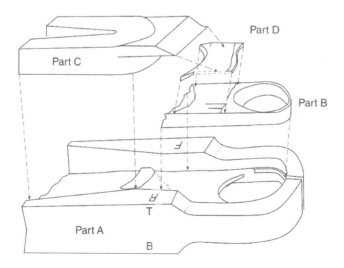

FIGURE 9.2 Failed module of assembly diagram.

TABLE 9.1

Annotation Used for Movement Diagram

S. No	Symbol	Meaning
1	+450/−450	Ply interface
2		Damage introduced during dissection
3	⟶	Direction of shear
4	⟶	Direction of crack growth

into two types, namely, mechanical and chemical damage. The overall procedure of postoperative damage sits is cleared the events of initial failure.

During the real failure event, they are familiar in the mechanical damage (Grove and Smith 1987). Compression crushing is an example of mechanical damage and outward crushing is constant. However, the post-failure mechanical damage should be avoided, which is induced by avoiding the surface of diluting. Most damage should avoid the confusion in the exam.

Most fracture is enclosed with the morphological debris which is visible to the fibre combination of contraction and tension. In post-failure damage, they complicate the images of the original fracture and it is difficult to interpret. Likewise, the dilute surface is obvious with the debris as well as crushing, as shown in Figure 9.3. During the lay-up, the example of the chemical damage is shown in Figure 9.4 (Greenhalgh 2009).

FIGURE 9.3 Post-failure mechanical damage to an interlaminar fracture.

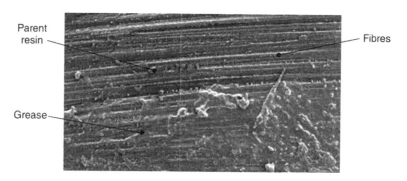

FIGURE 9.4   Grease contamination to a delamination fracture surface.

REFERENCES

Armstrong, K., Cole, W. and Bevan, G., 2005. Introduction to composites and care of composite parts. *Care and Repair of Advanced Composites.* Armstraong, Chesmar, Museux and Cole, SAE Publishers, pp. i–xxviii.

Becker, W.T., Shipley, R.J., Lampman, S.R., Sanders, B.R., Anton, G.J., Hrivnak, N., Kinson, J., Terman, C., Muldoon, K., Henry, S.D. and Scott Jr, W.W., 2002. *ASM Handbook.* Volume 11 Failure Analysis and Prevention. ASM International, p. 1072.

Berthelot, J.M., 1999. *Composite Materials. Mechanical Behaviour and Structures Analysis.* New York: Spring-Verlag, p. 620.

Bonhomme, J., Argüelles, A., Viña, J. and Viña, I., 2009. Fractography and failure mechanisms in static mode I and mode II delamination testing of unidirectional carbon reinforced composites. *Polymer Testing*, 28(6), pp. 612–617.

Dorworth, L.C., 2001. Composite tooling. In ASM Handbook, p. 21.

Greenhalgh, E. and Hiley, M., 2008. *Fractography of Polymer Composites: Current Status and Future Issues.* Faculty of Engineering Aeronautics, Stockholm, Sweden.

Greenhalgh, E., 2009. *Failure Analysis and Fractography of Polymer Composites.* Elsevier, Netherlands.

Grove, R.A. and Smith, B.W., 1987. *Compendium of Post-Failure Analysis Techniques for Composite Materials.* Seattle, WA: Boeing Military Airplane Co. USA

ASM International. 1987. *ASM Handbook*, Volume 12: Fractography. Materials Park, OH: ASM International, p. 517.

Kar, R.J., 1992. *Composite Failure Analysis Handbook.* Volume 2. Technical Handbook/Part 2. Atlas of Fractographs. Hawthorne, CA: Northrop Corp Hawthorne CA Aircraft Div.

Laš, V., Zemčík, R. and Měšťánek, P., 2006. Numerical simulation of composite delamination with the support of experiment. *Acta Mechanica Slovaca*, 10(1), pp. 303–308.

Lemascon, A., Castaing, P. and Mallard, H., 1996. Failure investigation of polymer and composite material structures in the mechanical engineering industry. *Materials Characterization*, 36(4–5), pp. 309–319.

McCoy, R.A., 2004. SEM fractography and failure analysis of nonmetallic materials. *Journal of Failure Analysis and Prevention*, 4(6), pp. 58–64.

Purslow, D., 1984. Composites fractography without an SEM—The failure analysis of a CFRP I-beam. *Composites*, 15(1), pp. 43–48.

Quinn, J.B. and Quinn, G.D., 2010. Material properties and fractography of an indirect dental resin composite. *Dental Materials*, 26(6), pp. 589–599.

10 Applications of Various Types of Polymer- Based Composites

Vigneshwaran Shanmugam
Saveetha Institute of Medical and Technical Sciences

N.B. Karthikbabu
Centurion University of Technology and Management

Sundarakannan Rajendran
Kalasalingam Academy of Research and Education

Oisik Das
Luleå University of Technology

CONTENTS

10.1 INTRODUCTION

Polymer composites have been used in many engineering as well as household applications as they have been strengthened with functional fillers and reinforcements. This feature makes them adaptable to different modern applications. Polymer composites are the main focus of design engineers in any lightweight engineering application due to their combination of strength and durability. With their superior structural properties, they give many benefits for certain uses relative to the usage of conventional materials. One of the main characteristics is their light weight while at the same time providing improved specific strength, which reduces overall weight up to 40%. In addition, the characteristics of polymer composites can be modified to provide high strength with low weight, corrosion resistance to certain chemicals, and to provide extremely resilient structures under the most adverse environmental conditions. Based on the type of application, polymers and their composites are categorized as structural and non-structural composites. Structural composites have increased strength and modulus which are used for load-bearing applications, whereas non-structural composites have relatively low strength and are not used for load-bearing applications. At present, polymer composites are used in many fields, such as aerospace, maritime, industrial, defense, building, and consumer goods. This chapter outlines the application of polymer composites in the current engineering environment.

10.2 FIBER COMPOSITES IN AUTOMOTIVE SECTORS

In the automotive industry, polymer composites are used to reduce the weight of vehicles, thereby increasing the efficiency and performance of vehicles. In particular, polymer composites are found in supercars and sports cars to boost performance in order to stay competitive in the automobile market. In automotive vehicles, polymer composites are found in both the interior and exterior parts, such as door panels, seat backs, headliners, package trays, dashboards, interior trim panels, bins, and energy absorbing composites crash elements. Polymer composites in automobiles also reduce fuel economy as it is estimated that about 75% of fuel consumption is attributed to vehicle weight. The use of polymer composites in vehicles can reduce the weight by 20%–40% and tooling costs by 40%–60% (Das 2001). In addition, assembly costs and part consolidation time can also be reduced. The energy absorbed by the composite structure during impact is higher than metallic structures. Composites have superior crash performance than traditional steel. Metallic structures undergo plastic deformation during impact, while composites absorb impact and undergo deformation, local cracking, and matrix deformation. Metals like steel can typically absorb about 35 J/kg of energy, but thermoset composites can absorb twice as much energy as steel.

Polymer composites offer increased resistance to corrosion, scratches, and dents in automobile parts. The application of polymer composites to automobiles involves three major components, such as body component, engine component, and chassis component. Compared to the engine and chassis component, the car body components need good impact resistance, strength, and ultra-fine finishing. Synthetic fibers such as glass and carbon fibers have been used to make body components for automobiles. Carbon fiber composites are an effective material than glass fiber composites for the production of automotive parts; however, glass fiber is economical in such a way that manufacturers are using glass fiber composites in automotive parts for a long period of time. However, carbon fiber composites are also used in some advanced vehicle designs. In most vehicle designs, E-glass-type glass fibers were selected for reinforcement, which may be in random orientation or mat form. In the random orientation type, short and discontinuous fibers were used with typical length varying from 20 to 50 mm. Both thermoplastic and thermoset plastic materials were preferred on the basis of the design needs of the components. Some thermoplastic fiber composite automobile components are dash boards, front ends, spare tires, holders, etc. The manufacturing methods varies based on the type of the matrix material used. For fabricating thermoplastic composites, fabrication techniques like extrusion, thermoforming, stamp forming process, and flow pressing process were used. Manufacturing methods such as the compression molding method, structural reaction injection molding, and resin molding method were preferred for the production of thermoset plastic automotive parts (Chung 2010). The automobile parts made from compression molding process are called sheet molding compounds (SMCs) (Boylan and Castro 2003). Typical examples of SMC components in automobiles include radiator supports, bumper beams, roof frames, door frames, engine valve covers, timing chain covers, oil pans, hoods, pickup boxes, deck lids, fenders, spoilers, and so on. The main drawback of SMC components is the poor surface finish. In case of body components, the surface finish must be in class A, which is unfortunately not feasible in SMC components. For about two centuries, SMC components have been used in the automotive industry, which has made significant changes in both production and performance. Honda used carbon fiber composites (CFRP) in the body panels including the fenders and quarter panels of the Acura NSX-sports car. The second-generation Honda Ridgeline used SMC and multi-composite construction materials to construct a truck bed. In 2016, Honda introduced fiber composites in the front end module of Honda Civic. Fiber composites are used in Volkswagen Atlas, Golf, Jetta, Passat, and Tiguan models from interior panels to engine bay components. The Volkswagen Modular Sportscar System introduced a hybrid aluminum/CFRP composite platform in 2014. The chassis component has been developed with 13% CFRP, which is 15% lighter than the aluminum platform. Using CFRP, a 40% increase in torsional rigidity and tremendous performance were achieved. The use of CFRP in the BMW i3 vehicle body has reduced production time by half compared to conventional automobiles which makes it more economical. The BMW i3 is made of a CFRP body and uses an exclusively designed 100% automated bonding technique instead of screws, rivets, and welding, which has reduced assembly time and cost compared to conventional metals. In addition, the use of CFRP in BMW i3 reduced painting costs. In the case of a metal body, unique paints have been used to prevent corrosion, which is eliminated in the CFRP body because they do not corrode. Components of Audi A8 such as rear seat shell and rear panel were developed using continuous carbon fiber composite. In Toyota's Lexus LFA, 65% of the cabin was made using CFRP. CFRP in LFA is made of three different molding processes:

- The main cabin frame was made through prepreg;
- Transmission tunnel, floor panel, roof, and hood were made using resin transfer molding process;
- C-pillar and the rear floor were fabricated as SMC.

The CFRP structure used in the LFA has been developed by the LFA team. The advantage of this methodology was improvement in the automated weaving process with laser thread sensors, which guarantees fabric integrity at reduced production time. A three-dimensional carbon fiber loom has been specifically developed for the LFA program, and the LFA Group has also developed a bonding method for bonding carbon fiber and metal parts. Customized joining methods use a threaded aluminum embed enclosed by the CFRP. However, the LFA uses a system that does not require any insert or direct contact with the CFRP using a flanged aluminum collar to link the two materials, thus overcoming the inherent weaknesses of such joints. Toyota

Prius PHV is perhaps the first mass-produced vehicle in the world to have CFRP-made tailgate. McLaren automotives used CFRP for fabricating chassis in high-end racing cars like MP4/1, 675LT, and 720S.

In addition to synthetic fiber composites, natural fiber-reinforced composites have also been used in certain specified parts of automobiles. Table 10.1 lists the natural fiber composite automobile components by different automobiles

TABLE 10.1

Applications of Natural Fiber Composites in Automobile Field

Manufacturer	Model	Components
Audi	A2, A3, A4, A4 Avant, A6,A8, Road star, Coupe	Audi seat back, side and back door panel, boot lining, hat rack, spare tire liner
BMW	3, 5, 7 series	Door panels, headliner panel, boot lining, seat back, noise insulation panels, molded foot well linings
Citroen	C5	Interior door paneling
Chrysler	Chrysler Sebring	Interior door panel
Daimler-Benz	Mercedes A, C, E, S class, Trucks EvoBus (exterior)	Door panels, windshield/dashboard, business table, piller cover panel, glove box, instrumental panel support, insulation, molding rod/apertures, seat backrest panel, trunk panel, seat surface/backrest, internal engine cover, engine insulation, sun visor, bumper, wheelbox, roof cover
Fiat	Punto, Brava, Marea, Alfa Romeo 146, 156	Door panel
Ford	Mondeo CD 162, Focus, freestar and Lincoln MKZ	Floor trays, door panels, B-piller, boot liner, seating headrests
General Motors	Cadillac Deville, Chevrolet, TrailBlazer	Seat backs, cargo area floor
Honda	Pilot	Cargo area
Lotus	Eco Elise	Body panels, spoiler, seats, interior carpets
Nissan	Nissan Leaf	Floor mats
Mitsubishi	Space star, Colt	Cargo area floor, door panels, instrumental panels
Opel	Astra, Vectra, Zafira	Instrumental panel, headliner panel, door panels, Pillar cover panel
Peugeot	406	Front and rear door panels
Renault	Clio, Twingo	Rear parcel shelf
Rover	2000 and others	Insulation, rear storage shelf/panel
Saturn	L3000	Package trays and door panel
Toyota	Raum, Brevis, Harrier, Celsior	Door panels, seat backs, floor mats, spare tier cover
Volkswagen	Golf A4, Passat, Variant, Bora	Door panel, seat back, boot-lid finish panel, bootliner
Volvo	C70, V70	Seat padding, natural foams, cargo floor tray

Source: Ngo, Tri-Dung 2018.

manufacturers. In 2006, EU legislation was implemented which signifies the importance of natural fiber composites in automobile applications. According to the legislation, 80% of car parts should be recyclable and reusable (Ferrao and Amaral 2006). Although synthetic fiber composites, such as fiberglass and carbon fiber composites, are proven to be efficient materials for automotive applications, they are not recyclable and reusable and are hazardous. Natural fiber composites are lighter than glass fiber composites, have low density, are cost-effective, environmentally friendly, renewable, and non-hazardous. Leading automotive manufacturers such as BMW, Volkswagen, Mercedes Benz, Audi, Toyota, and Ford use natural fiber composites in their automotive parts. In 1941, the body of Ford vehicles was made by soyabean and hemp fiber composites. In 2014, Ford used cellulose-based polypropylene composites in the arm rest of the 2014 Lincoln MKX. Rice hulls-reinforced plastics are found in F-150 pickup truck wiring harnesses. Wheat straw-reinforced plastics are used in Flex full-size SUV storage bins. Ford Mondeo pointed out in 1996 that the internal car boards were constructed of kenaf-based composites. Ford uses natural fiber composites as a substitute to glass fiber composites and claims that the material is 10% lighter, increasing performance by 30% and minimizing carbon emissions. BMW used natural fibers composites in the 3, 5, and 7 series vehicle parts such as seat backs, doors panels, and boot linings. In 2001 almost 4,000 t of natural fiber was used by BMW in 3 series alone. Flax and kenaf fiber-based polymer composites were used in door panels and package trays of the Saturn L300 and Opel Vectra vehicles as synthetic fiber replacements. Similarly, a Malaysian local automotive firm fabricated seat backs, package trays, door panels, and headliners for the Perodua and Proton car models using kenaf-based polymer composites.

10.3 FIBER COMPOSITES IN AEROSPACE SECTORS

The exploration of research and development in the aerospace industry to increase the performance of commercial and military aircraft have led to the development of high-performance structural materials. Polymer composites have been used in the aerospace industry due to their light weight and increased performance. Moreover, they have excellent physical characteristics, low density, and high stiffness. Fiber composites made of glass and carbon fibers have been used in commercial transport aircraft, military fighter aircraft, helicopters, satellites, and missiles. In aerospace industries, composites are used in engine blades, brackets, interiors, nacelles, propellers/rotors, single-aisle wings, and wide body wings (Kumar and Padture 2018). Although only a small percentage of polymer fiber composites contribute to the total weight of aircraft, these materials are found to be sophisticated in aircraft applications. The key aspects in material selection for aerospace applications are light weight, high strength, high stiffness, and good fatigue resistance. Composites have been used in aerospace applications for a long time in non-safety critical aircraft applications, but in recent decades, composites have been applied to primary aircraft structures such as wing structures, engine components, and fuselage found in modern Boeing, Airbus, and Bombardier aircrafts. General Dynamics manufactures lightweight and high-performance composite components for various aircraft applications, such as wing and fuselage structures, flight control surfaces, doors, engine

TABLE 10.2
Composite Made Components in Aircraft

Model	Components
Wing Structures	Edge beams, flap leading and trailing edges, flap track fairings, wing panels, and winglets
Fuselage Structures	Canopy fairings, pylons, spars, and vertical tail skins
Doors	Avionics, baggage, emergency exit, landing gear, main entrance, and service and weapons bay
Flight Control Surfaces	Rudders, elevators, and flaps

structures, and pressure vessels. Table 10.2 shows aircraft components made of composites by General Dynamics.

In Airbus 320, application of composites in various components reduced weight by 800 kg. Airbus A380 comprises 16% of composites in its structural weight, the center wing box in A380 weighs about 8.8 t, of which 5.3 t is the weight of composite materials. Figure 10.1 shows the different composite components used in aerospace applications. Boeing 787 airframe is made of carbon fiber composites and other composites, which saves 20% of the weight compared to conventional aluminum airframe. Approximately 32 t of CFRP are used in Boeing 787 aircraft, where carbon fiber alone weighs 23 t. Addition of composites to the airframe offers increased toughness and reduced weight, which can also reduce the maintenance costs of the aircraft. The composites do not rust or corrode, which reduces future maintenance costs. A350

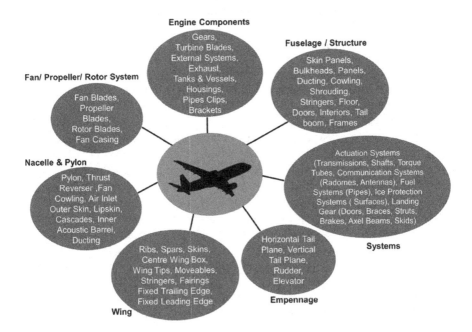

FIGURE 10.1 Composite components in aerospace applications.

TABLE 10.3

Composite Components in Airbus A310 and A320

Composite Components in Airbus A310	Composite Components in Airbus A320
Vertical and horizontal stabilizer, flap fairing, motor most, air conditioning components, front landing gear hatch, karman, radome, main landing gear hatch and fairing, disk brakes, flap, ailerons, spoilers, and rudder	Leading and trailing edge of vertical stabilizer, horizontal stabilizer, flap rails and cover fairing, cabin component, landing gear hatch, karman, radome, main landing gear hatch and cover, motor case, disk brakes, floor, external ailerons, external spoilers, internal spoilers, and rudder

XWB requires 50% less structural maintenance tasks. For aircrafts with conventional metals, the threshold for airframe checks is 12 years, but for A380, it is 8 years due to the increased use of composites in its total weight percentage. Various composite components used in Airbus A310 and A320 are listed in Table 10.3. In aircrafts, corrosion and fatigue in structures are critical, which significantly increase non-routine maintenance. Non-routine maintenance may lead to an increase in the total working hours of maintenance checks. Expanded use of fiber composites in Boeing 787 reduced non-routine labor costs compared to the conventional metallic airframe. Glass fiber-reinforced aluminum (GLARE) was widely used in the upper fuselage skins of A380; GLARE contributed 3% of the total weight of the A380 structure. Glare can contribute 15%–30% weight savings over aluminum and exhibit outstanding fatigue properties. Among combat aircrafts, Harrier II was the first aircraft with extensive use of carbon fiber composites. In total weight of Harrier, 26% of the weight is shared by composites, which reduces the weight by 217 kg compared to the conventional metallic structure. Composites are used in the components of fuselage and rotor blades in helicopters. In commercial and military helicopters, composites made of glass and carbon fiber have been used in the body and tail boom, which are effective in reducing weight, vibration, and corrosion. The increased fatigue resistance of the fiber composites may increase the life of the rotor blades. In Bell V-280 Valor, carbon fiber composites are used in the fuselage, wing, and tail in the form of honeycomb sandwich configurations resulting in a weight saving of 30%.

10.4 FIBER COMPOSITES IN CONSTRUCTION

Polymer-based composites have been successfully used in the construction industries for tank liners, roofs, pressure pipes, load-bearing, and infill panels. In recent years, fiber composites have been used in road construction where they are capable of eliminating deterioration problems in highways and other concrete structures. In the construction of concrete structures, the fiberglass rebar replaces heavy and corrosive conventional steel. The fiber glass rebar has a tensile strength two times higher than that of steel that offers a long service life. The light weight and increased strength of the fiber glass rebar reduce the total weight of the structure and facilitate the bridge to withstand live loads (Deitz et al. 2003). In addition, the durability and corrosion resistance characteristics of the fiber glass rebar improve the life of the bridges. The corrosion resistance of the fiber glass rebar enables the structure to resist corrosion

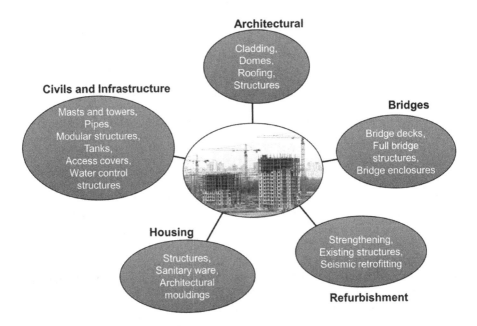

FIGURE 10.2 Composites in construction applications.

due to dynamic environmental weathering. TUF-BAR is one of the pioneers in the manufacturing of rebar fiber glass. Figure 10.2 shows the application of fiber composites in the construction industries. Fiber composites offer good esthetic appearance and can be used for linings, fittings, and cladding. In architectural applications, fiber composites are used to make sculptures, facades, domes, and clock towers that reduce weight, time of installation and maintenance, and further enhance durability.

10.5 POLYMER AND ITS COMPOSITES IN ELECTRICAL AND ELECTRONICS APPLICATIONS

Polymers such as thermosets, thermoplastics, and their composites have been used in the electrical industry due to the versatility of polymer properties. Polymers and composites used in electrical applications critically have high/low thermal conductivity, low thermal expansion, and high/low electrical conductivity, depending on the type of application (Pathania and Singh 2009). Polymer and its composites are used in electrical and electronic applications in the form of thin films and as coating materials. In electrical and electronic applications, polymer and its composites have been used as electrical insulating material, electrically conductive material, and as adhesives, coatings, potting compounds, and sealants. For electrical applications like wires and cables, flexible polymers and its composites have been used, and for making enclosures and covers, rigid polymers have been used which should have good modulus, strength, elongation, impact resistance, fire resistance, thermal resistance, anti-static performance, and durability. Figure 10.3 shows the properties needed for the rigid polymer and its composites for electrical applications. Further environmental factors

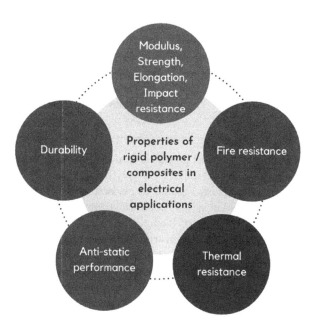

FIGURE 10.3 Properties of rigid polymer and its composites in electrical applications.

are critical to be considered while selecting the material as it affects the durability of the polymer. Because electrical and electronic devices are universal devices that are used in hot and cold environments such as deserts, rainy areas, and extreme climatic conditions. Polymers and composites are also used in outer space environments such as satellites. Fire resistance is other crucial factor to be considered in the polymer for electrical applications, such as the use of PVC wires and cables, which are good fire-resistant materials. Some applications of polymer and its composites in electrical and electronics are printed circuit boards (PCBs), PCB construction, embedding of printed circuits, polymer thick films, encapsulation of integrated circuits, cermet thick films, solder pastes, dielectric tape, conductive adhesives, non-conductive adhesives, high-temperature applications, electromagnetic interference (EMI)/radio-frequency interference (RFI) shielding, electrostatic dispersants, and pressure sensors. In addition, thermal conductivity-enhanced polymer and its composites are widely used in microelectronic devices for their promising heat transfer ability.

10.6 BIOPOLYMER/BIO-NANOCOMPOSITES AND ITS APPLICATIONS

Biocomposites are developed from renewable natural sources and are used in various applications that reduce dependence on fossil fuels and develop an eco-friendly environment. Research and development in bio-nanocomposites opened the door to the development of high-performance biodegradable materials for various applications that can replace the conventional non-biodegradable petroleum-based materials.

TABLE 10.4
Biopolymer Types

Chemical Synthesis	Biological Resources	Microorganisms
From Biomass – PLA	Lipid – Wax and fatty acids	Polyester – PHAs
From Petrochemicals –	Protein – Gelatin, gluten, cornzein, etc	(PHB, PHBV)
PCl, PVA, PGA	Carbohydrates – Starch, cellulose, chitosan,	Carbohydrates –
	agar, etc	Pullulan, Curdlan

Biopolymers such as polylactic acid (PLA), polyhydroxyalcanoates (bacterial polyesters), polycaprolactone (PCL), poly(butylene succinate) (PBS), polyhydroxybutyrate (PHB), starch plastics, cellulose plastics, soy-based plastics, etc., and their composites have been used in various commercial applications (Sadasivuni et al. 2020). Table 10.4 shows various kinds of biopolymers. These biopolymers and composites are used in paper and packaging industry, electronics industry, painting, coating, and biomedical applications. Biopolymer nanocomposites are used in biomedical applications in scaffold tissue engineering, medical implants, drug release, gene delivery, dental applications, wound healing, and so on. Polymers used in packaging applications should have good mechanical, thermal, anti-microbial, and biodegradable properties.

10.7 ROLE OF POLYMER COMPOSITES IN BIOMEDICAL APPLICATIONS

Polymer and its composites are extensively used in biomedical applications such as drug delivery, dental, implant material, tissue engineering, and regenerative medicine applications. For instance, PLA has evolved as an attractive polymeric material for biomedical applications due to its properties such as biocompatibility, biodegradability, mechanical properties, and process capability. In addition, the thermoset type of bisphenyl-polymer/CRFP composites are considered as advanced materials to replace metal bone implants. CFRP composite can offer densities and electrical conductivity/resistivity properties close to bone with strengths much more than metals on a per-weight basis. Silver (Ag) nanoparticles are well known for their tendency to antibacterial activity. In addition to thermal and mechanical properties, the antibacterial activity is significantly improved when low concentration of Ag nanoparticles is reinforced in polymer. Cytotoxicity of Ag nanoparticles is reduced when they are embedded in a polymer matrix. In orthopedic applications, composite materials are widely used as bone graft, bone fracture internal fixation devices, joint prostheses (such as hip, knee, bone cement, and so on), artificial tendons, artificial ligaments, and artificial cartilage applications. In practice, ultra-high-molecular-weight polyethylene (UHMWPE) and its composites are potentially used material for the acetabular cup of artificial hip joints due to its superior physical and mechanical properties compared to other polymers (Michael Sobieraj and Marwin 2018). UHMWPE has been used extensively in joint implants for more than four decades, especially as joint filler in total hip replacements and tibial inserts in total knee replacements. Hence, attractive features like low weight, high specific strength, and low cost of the polymer and its composites make them a potential candidate in biomedical applications.

REFERENCES

Boylan, S. and Castro, J.M. 2003. Effect of reinforcement type and length on physical properties, surface quality, and cycle time for sheet molding compound (SMC) compression molded parts. *Journal of Applied Polymer Science*, 90(9): 2557–2571.

Chung, D.L. 2010. *Composite Materials: Science and Applications*. Germany: Springer.

Das, S. 2001. The cost of automotive polymer composites: A review and assessment of doe's lightweight materials composites research. United States. doi:10.2172/777656.

Deitz, D.H., Harik, I.E. and Gesund, H. 2003. Physical properties of glass fiber reinforced polymer rebars in compression. *Journal of Composites for Construction* 7: 363–366.

Ferrao, P. and Amaral, J., 2006. Assessing the economics of auto recycling activities in relation to European Union Directive on end of life vehicles. *Technological Forecasting and Social Change* 73(3): 277–289.

Kumar, S. and Padture, N.P. 2018. Materials in the aircraft industry. In Kaufman, B. and Briant, C. (eds.), *Metallurgical Design and Industry* (pp. 271–346). Cham: Springer.

Michael Sobieraj, M.D. and Marwin, S. 2018. Ultra-high-molecular-weight polyethylene (UHMWPE) in total joint arthroplasty. *Bulletin of the NYU Hospital for Joint Diseases* 76(1): 38–46.

Pathania, D. and Singh, D. 2009. A review on electrical properties of fiber reinforced polymer composites. *International Journal of Theoretical & Applied Sciences* 1(2): 34–37.

Sadasivuni, K.K., Saha, P., Adhikari, J., Deshmukh, K., Ahamed, M.B. and Cabibihan, J.J. 2020. Recent advances in mechanical properties of biopolymer composites: A review. *Polymer Composites* 41(1): 32–59.

Index

Printed in the United States
by Baker & Taylor Publisher Services